THE BIG IDEAS IN PHYS
AND HOW TO TEACH THEM

M000305569

'The book is brilliant. I hope all physics teacher trainers and trainees, as well as established teachers, use this critically important work to guide their teaching.'

John Sweller, Emeritus Professor at the School of Education,
The University of New South Wales, Australia

The Big Ideas in Physics and How to Teach Them provides all of the knowledge and skills you need to teach physics effectively at secondary level. Each chapter provides the historical narrative behind a Big Idea, explaining its significance, the key figures behind it and its place in scientific history. Accompanied by detailed, ready-to-use lesson plans and classroom activities, the book expertly fuses the 'what to teach' and the 'how to teach it', creating an invaluable resource which contains not only a thorough explanation of physics, but also the applied pedagogy to ensure its effective translation to students in the classroom.

Including a wide range of teaching strategies, archetypal assessment questions and model answers, the book tackles misconceptions and offers succinct and simple explanations of complex topics. Each of the five big ideas in physics are covered in detail:

- electricity
- forces
- energy
- particles
- the universe

Aimed at new and trainee physics teachers, particularly non-specialists, this book provides the knowledge and skills you need to teach physics successfully at secondary level, and will inject new life into your physics teaching.

Ben Rogers teaches physics and trains new teachers for Paradigm Trust. He is a former lecturer on the Physics Enhancement Course at the University of East London, UK.

THE BIG IDEAS IN PHYSICS AND HOW TO TEACH THEM

Teaching Physics 11–18

Ben Rogers

Routledge
Taylor & Francis Group

LONDON AND NEW YORK

First published 2018
by Routledge
2 Park Square, Milton Park, Abingdon, Oxon OX14 4RN

and by Routledge
711 Third Avenue, New York, NY 10017

Routledge is an imprint of the Taylor & Francis Group, an informa business

British Library Cataloguing-in-Publication Data
A catalogue record for this book is available from the British Library

Library of Congress Cataloging-in-Publication Data
Names: Rogers, Ben (Physics writer), author.
Title: Big ideas in physics and how to teach them: teaching physics 11-18 / Ben Rogers.
Description: Abingdon, Oxon: Routledge, 2018.
Identifiers: LCCN 2017059174 (print) | LCCN 2018002921 (ebook) | ISBN 9781315305431 (ebook) | ISBN 9781138235076 (hardback) | ISBN 9781138235069 (pbk.)
Subjects: LCSH: Physics teachers–Training of. | Physics–Study and teaching (Secondary) | Physics–Study and teaching–Activity programs.
Classification: LCC QC30 (ebook) | LCC QC30 .R635 2018 (print) | DDC 530.071–dc23
LC record available at https://lccn.loc.gov/2017059174

ISBN: 978-1-138-23507-6 (hbk)
ISBN: 978-1-138-23506-9 (pbk)
ISBN: 978-1-315-30543-1 (ebk)

Typeset in Interstate
by Deanta Global Publishing Services, Chennai, India

For Denise, Laura and Hannah

CONTENTS

Preface   *x*

Acknowledgements   *xi*

Introduction   1

Zero   A big idea about learning   3

Working memory 3

Long-term memory 4

External environment 4

How can we use Cognitive Load Theory to accelerate learning? 5

Knowledge 10

Archetypal questions 10

Model-based problem solving 11

The privileged status of stories - Willingham 11

Misconceptions - When knowledge is wrong 11

Practical work in physics 13

Reducing Cognitive Load for practical work 13

Literacy - A different sort of physics problem 14

What are the Cognitive Loads associated with reading and how can we reduce them? 14

What are the Cognitive Loads of writing and how can we reduce them? 16

How to teach writing in physics 18

Conclusion 18

1 Electricity   20

Introduction 20

A history of electricity 20

Electricity in the Classroom 36

Misconceptions 38

Archetypal questions 40

Models 40

Model based reasoning 41

Practical electricity   43
Example lesson plan   44
Conclusion   47

2 **Forces at a distance**   49
Petrus Peregrinus, Crusader – 1269   50
William Gilbert of Colchester, Physician to Queen Elizabeth I – 1600   50
Newton's Law of Universal Gravitation – 1687   51
Faraday's lines of force – 1852   53
Maxwell's equations: The second great unification in physics – 1865   54
Einstein's curved space – 1915   55
Fermi's nuclear forces – 1933   55
Teaching forces at a distance   56
Archetypal questions   58
Using strategies from cognitive psychology in lessons   59
Using demonstrations and practical work for writing   61
Example lesson plan   63
Conclusion   67

3 **Energy**   69
A short history of five energies   69
Kinetic energy and potential energy: Descartes and Leibniz – 1644
   and 1676   70
Chemical energy and heat energy: James Joule – 1843   71
Nuclear energy: $E = mc^2$ – 1905   72
Teaching energy   73
Types of energy – stores and pathways   73
Misconceptions   73
Archetypal questions   74
Using strategies from cognitive psychology in lessons   76
Reading and writing   76
Reducing Cognitive Load   77
Example lesson plan   77
Conclusion   85

4 **Particles**   87
Introduction   87
A history of particles   87
But atoms are not real. Or are they? Einstein – 1904   89
Rays, beams and other phenomena – 1869 to 1899   91
Pieces of atoms – 1897 to 1899   92
Disproof of the pudding: Rutherford's astonishing career – 1900
   to 1921   93
Neutrons and war – 1932 to 1945   95
Teaching particles   96

Misconceptions   96
Archetypal questions   97
Useful strategies from cognitive psychology in lessons   99
Example lesson plan   101
Conclusion   105
Notes   105

5 The universe   **107**
Introduction   107
The telescope – 1608   110
Teaching the universe   115
Misconceptions   115
Archetypal questions   117
Models   118
Practical astronomy   123
Example lesson plan   126
Conclusion   129

Bibliography   *130*
Index   *133*

PREFACE

This book is written for every new physics teacher, whether you are new to teaching or new to teaching physics. Recruiting new physics teachers is difficult. In 2015/16 in England:

- 29% of the 1,055 physics training places were unfilled.
- 28% of physics lessons were taught by teachers without post A-level experience.

(DfE 2016)

The initial idea for the structure of this book came from a report published in 2010 by the Association of Science Education (ASE): 'The Principles and Big Ideas of Science Education', edited by Wynne Harlen. The report identified fourteen 'Big Ideas' of science education. I have fewer *big ideas* for physics teachers. The ideas I have chosen are *electricity, forces at a distance, energy, particles* and *the universe*. Each of these big ideas has its own stories and its own pedagogy. This book has a chapter for each.

But the book starts with a different sort of big idea: a learning theory. I have used John Sweller's Cognitive Load Theory to explain why I have chosen specific activities and approaches.

Whether you are an experienced teacher, teaching physics for the first time or new to the profession, thank you. My aim is to help you enjoy teaching physics and to teach it well.

ACKNOWLEDGEMENTS

This book began as a physics knowledge enhancement course at Thetford Academy. I got to work with enthusiastic, skilled teachers who were teaching physics but who weren't experienced physics teachers. I want to thank Adrian Ball for initiating the course and for encouraging me to write the book, and all of the participants of the course for their encouragement and feedback.

The narrative element of the book was developed from a discussion with Daisy Christodoulou about the importance of narrative in learning. We wanted a text which told the narratives of science, something like Bill Bryson's *A Short History of Nearly Everything* but aligned with the curriculum. Half of this book is dedicated to those narratives.

I would like to thank Alan Weller of UEL for guiding me through the early stages of finding a publisher and Bill Holledge of Paradigm Trust, Dr Jo Saxton of Turner Schools and Tony Sherborne of the Centre for Science Education, Sheffield Hallam University for helping me clear the first hurdles.

I am grateful for generous feedback on the first chapter from Professor Dylan Wiliam and Professor John Sweller. The errors which remain are my own.

My thanks also to Dr Lucy Rogers, Richard Heald and Simon Laycock for their feedback on the writing and their encouragement.

Finally, I owe so much in my life and career to Denise Dickinson, who has been by my side since I first learnt to teach, guiding and encouraging me and along the way teaching me to write a little better.

INTRODUCTION

Know how to solve every problem that has been solved.

Richard Feynman

You are a new physics teacher – you have been asked to teach students how to be physicists. This means teaching students how to become physics problem solvers.

A physicist is the sum of the problems she can solve. She knows the conservation of energy when she can solve all of the problems associated with it. Knowing all the problems lets you solve new ones:

> the science student, confronted with a problem, seeks to see it as like one or more of the exemplary problems he has encountered before.
>
> (Kuhn 1977: 297)

In other words, to become a better problem solver, a novice physicist needs to be exposed to as many archetypal questions as possible. More than that, she needs to be exposed to archetypal questions in as many guises as possible, until she can see the underlying deep structure of a question.

This is a book about solving physics problems. It is about the knowledge a learner needs to become an expert. It is about the archetypal problems every physics student needs to learn. It is about how to teach them as efficiently and effectively as possible.

In 1966 Richard Feynman gave an interview about teaching physics. He said that there is usually a problem in physics lessons – the students do not know where they are. His solution: "there always should be some kind of a map" (Feynman 2010: 16).

I have constructed this book around a map. I started by writing the stories around a time-line for five big ideas of physics: electricity, forces at a distance, energy, particles and the universe. Figure i.1 is a map I made of the big ideas of physics.

This book is about teaching these five big ideas. Each chapter starts with the story of the big idea. Stories find a way of lodging in our brains. I use the stories as a base to build knowledge onto. But before I start with the big ideas of physics, I need to explain about another big idea in this book: a big idea about learning.

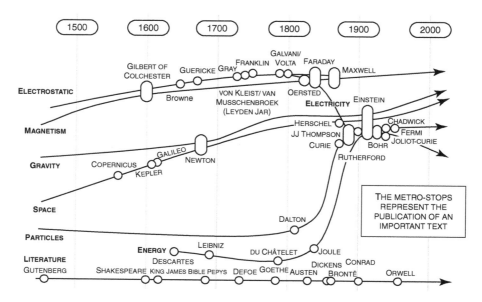

Figure i.1 A series of timelines showing when key events in the history of physics took place

Zero   A big idea about learning

This idea is Cognitive Load Theory (CLT), which has been slowly gaining recognition since it was first developed by John Sweller in the 1980s. I have used the theory throughout this book to recommend activities and strategies and to explain why they work. CLT is not a theory-of-everything, but it helps explain how we learn to solve problems.

CLT emphasises two types of memory: working memory and long-term memory, how they interact with each other and the external environment, as shown in Figure 0.1.

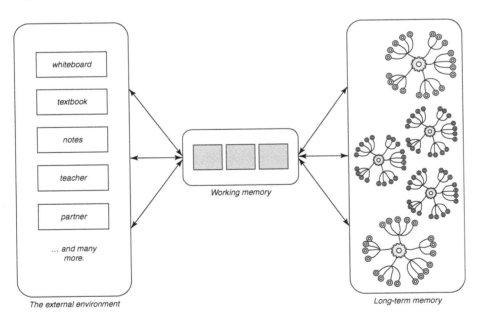

Figure 0.1 A model illustrating the external environment, working memory and long-term memory

Working memory

When we solve a problem, we store relevant, temporary information in our working memories. It is our mental workbench. Typical information manipulated in our working memories includes:

- data to solve the task;
- information about the task;
- relevant processes and strategies;
- information about social interactions (I use group-learning a lot in my classes).

The first important thing for teachers to know is that working memory is easily overloaded when dealing with novel information. Keeping track of three ideas is usually too much. If you are asking students to carry out a novel task, collaborating with new people using knowledge they haven't memorised, and hope that they will be able to reflect on their learning at the same time, you may be out of working memory and out of luck.

The second thing to know is that solving a problem at the limit of her ability does not allow a student to reflect on the process - a vital part of learning. So, even if the learner gets the problem right, you may have wasted learning time.

Long-term memory

The reason that some of us are better at solving problems than others is mainly due to long-term memory. If you have two capable students solving a problem, the one with the most relevant knowledge in her long-term memory is more likely to be successful. Relevant knowledge includes:

- subject knowledge - facts and the relationships between facts.
- learning knowledge - how to access the knowledge we need to solve a problem.
- general knowledge - the information we assume others know. This is important for effective communication, especially when examples or questions are set within a context; for example, an examiner, textbook writer or teacher may assume the student knows about Africa, audio cassettes, the Tudors or snow.
- social knowledge - understanding the relationships, roles, rules and expectations of those around us. Social knowledge is key to solving problems in the real world.

We store our long-term memories as schemata - networks of knowledge organised in meaningful ways. To become better problem solvers, students need richer schemata with more knowledge, connected more meaningfully. For example, a student with a well-developed schema about pressure will know the effects of pressure on: living things; elephant feet, high-heeled shoes, snow-shoes and drawing pins; submarines and high-altitude balloons; the bends; early vacuum pumps and the Magdeburg hemispheres; the gas laws (Boyle's, Gay-Lussac's and Charles') and their history; Boltzmann and the kinetic theory of gases; Brownian motion; and a bunch of equations involving pressure.

External environment

Our working memories can only take in a small amount of information from the external environment at one time, which then rapidly fades. This is important for teachers because we influence our students' external environment: the whiteboard, a demonstration, a worksheet, the classroom display, the seating plan. The aim is to focus attention on relevant information and reduce distractions.

Learners also influence their external environments. When a task gets tricky and working memory gets strained, effective learners use their external environment to reduce the load on their working memories. That's why we count on fingers, do calculations on the backs of envelopes and work with others (making use of their working memories too). Every photo of a physicist shows a blackboard full of indecipherable, chalky marks – the classic useful external environment. Marking-up texts and diagrams, making notes, answering questions neatly and working effectively with peers are important strategies for learning.

How can we use Cognitive Load Theory to accelerate learning?

The first lesson from CLT is that students who know more, who have better developed schemata, are better at solving problems. If your primary goal as a physics teacher is to teach your students to be better problems solvers, your primary strategy has to be teaching your students more knowledge.

The second lesson is that reducing Cognitive Load makes learning more effective. When you reduce the Cognitive Load, learners solve problems more effectively and learn more. Reducing Cognitive Load means stripping out all of the extraneous, confusing detail and distractions from the task – especially for novices. Decide what you want your students to learn from a task and simplify everything else.

When students learn, they are developing schemata – the knowledge and organisation of knowledge in long-term memory. Our main job as teachers is to boost students' schemata. This means thinking hard about what we want our students to be able to recall instantly and with little effort – in other words: *knowledge*.

Sweller's research since the 1980s has shown that decreasing Cognitive Load increases learning. Researchers have found many effective strategies for reducing Cognitive Load to improve learning. I have described four strategies below: worked examples, completion problems, the goal-free strategy and reducing the split-attention effect.

Worked examples and completion problems

The problem with problem solving is that you need to be pretty knowledgeable before you become good at it. We tend to teach new information and then immediately put it into a problem. This doesn't help most learners.

CLT researchers have shown that an effective way to teach problem solving is by using worked examples. When a teacher models how to solve a problem, she is giving the guidance that novice physicists need. It is a way in: she makes the hidden process of solving the problem visible.

A worked example

A ball bearing falls through oil. The arrows in Figure 0.2 represent the forces acting on the ball.

Explain, in terms of forces, why the ball reaches a terminal velocity.

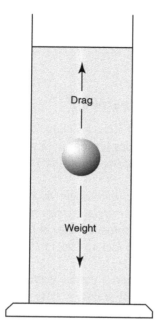

Figure 0.2 The forces on a sphere in a tube of viscous liquid

Model answer

Imagine you are standing at the board – ideally the question is projected adjacent to where you are explaining and making notes for the class:

1   The weight of the ball is independent of the ball's velocity – it doesn't change.
2   The drag on the ball increases as the ball accelerates (the drag more than doubles every time the velocity doubles).
3   The ball stops accelerating when the drag matches the weight – it has reached terminal velocity.

Try this several times with different contexts: a mouse falling down a well, a teacher jumping from a balloon, a small meteorite falling from space. Same concept, different context.

But then what? The jump from seeing someone do it to being able to do it yourself is still big and working memory is quickly overloaded.

One method is to give learners partially completed problems – this method is called *problem completion*. You reduce the cognitive load, allowing the learner to focus her working memory on fewer aspects of the problem.

Completion problems

Going straight from worked example to whole questions is very challenging for most learners. Completion problems are half-completed answers which focus the learner's attention on one element of solving the problem. The example below focuses a learner's attention on putting the correct values into an equation.

> *A 5kg ball rolls down a slope which is 2m higher at the top than at the bottom. How much more energy is in the ball's gravitational store at the top of the slope than at the bottom?*

$$m \ = \ \underline{\qquad}$$
$$h \ = \ \underline{\qquad}$$
$$g \ = 10m/s^2$$
$$E_g \ = mgh$$
$$\ = \ \underline{\qquad}$$
$$\ = \ \underline{\qquad} J$$

As a student masters each element of the problem, the support should be reduced.

For written answers, sentence starters reduce the cognitive load:

> *On 14 October 2012, Felix Baumgartner created a new world record when he jumped from a stationary balloon at a height of 39km above the Earth's surface. 42s after jumping, he reached a terminal velocity of 373m/s. Explain in terms of weight and drag how terminal velocity is reached.*

1   The weight _____.
2   The drag _____.
3   When the drag has increased _____.

One completion problem will not be enough. You will need lots. There are plenty of full questions available in past exam papers or you can make up your own. Your job is to take a full question and partially model the answer, leaving only the stages you want your students to practice. For example:

> *When his balloon experiment began to go wrong, Mr Rogers knew he had to jump. He was 5km high. Explain in terms of weight and drag why he reached terminal velocity as he fell.*

I have written sentence starters so that the learner does not have to sequence the answer for herself. Her task is to practice the individual stages of the answer.

1   His weight _____.
2   The drag _____.
3   When the drag has increased _____.

Completion problems are an effective method for focusing attention on specific elements of a problem. They reduce Cognitive Load by zooming in.

Another method of reducing Cognitive Load is to remove the question all together. This method is known as *goal-free*.

Reducing Cognitive Load by going goal-free

This strategy appears counter-intuitive, until you think about what you really want your students to learn. Figure 0.3 is a good example.

A plank of wood of mass 30kg and length 4.0m is used as a bridge across a muddy ditch. The plank is supported by the bank and a triangular support 1.0m from the other end of the plank as shown in the diagram.

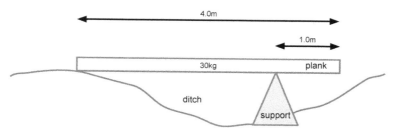

Calculate the vertical force acting on the plank from the triangular support.

Figure 0.3 A typical moments question

When you use this question in class, which of the following learning goals is most important to you:

A: learning how to solve this type of problem

or

B: finding out how much vertical force the support really supplies.

I'm assuming you chose A (like so many of the questions we set in physics classes, we don't really care about the answer to the question). Reducing the Cognitive Load allows the learner to learn. In the question above, simply cut out the text. You then have a situation to explore with your students - the plank on the support, as shown in Figure 0.4.

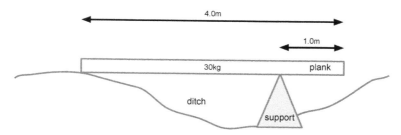

Figure 0.4 A typical moments question with the question text removed

I use a cooperative strategy at this stage, asking students to discuss the situation in pairs. This strategy is called *think-pair-share*. Each student has a copy of the diagram and makes as many annotations as they can in one minute. Then, for one minute they compare their annotations with a partner. While they are doing this, I walk around the class and choose two or three students to contribute particularly useful ideas to the whole class.

A considerable amount of learning has happened by this stage. They have practised retrieval, reinforcing their existing knowledge, and added to their *weights-on-beams* schema.

Your students may now be ready to tackle the problem. You may wish to demonstrate the worked example yourself or set a partially completed problem.

Going goal-free might sound directionless, but it is a powerful strategy for learning what can and cannot be done when faced with these given variables, leading to better problem solving.

Reducing the split-attention effect to reduce Cognitive Load

This strategy is about text and diagrams. When we have to split our attention between visuals and text, the Cognitive Load increases. How can you integrate the text into the question to reduce Cognitive Load?

Because the text is separate from the diagram, and quite wordy, the learner's attention is split, adding to the Cognitive Load for Figure 0.5. This reduces the students' ability to learn from the experience.

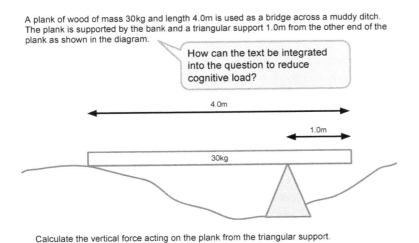

A plank of wood of mass 30kg and length 4.0m is used as a bridge across a muddy ditch. The plank is supported by the bank and a triangular support 1.0m from the other end of the plank as shown in the diagram.

How can the text be integrated into the question to reduce cognitive load?

4.0m

1.0m

30kg

Calculate the vertical force acting on the plank from the triangular support.

Figure 0.5 A typical moments question demonstrating the split attention effect

In Figure 0.6, I have adapted the question to minimise the *split-attention effect*. This leaves more working memory available for processing. Embedding the text in the image does more than reduce Cognitive Load; it uses a strategy with shown learning benefits called *dual-coding* (see Sumeracki and Weinstein (n.d.) (http://www.learningscientists.org/dual-coding/).

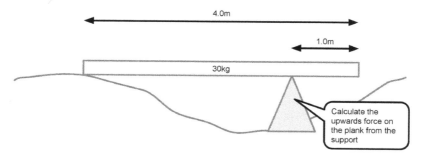

Figure 0.6 A typical moments question adapted to reduce the split attention effect

In all of these strategies, the aim is to reduce this support until your students can solve the problems on their own. In fact, when you continue to support for too long, Cognitive Load begins to increase again as the learner works around the support – this is called the *expertise-reversal effect*.

Knowledge

When Twitter arguments erupt over knowledge-based curricula in history or the canon in English literature, physicists scratch their heads. There is very little disagreement over what knowledge is important in the physics curriculum. There may be a disagreement over when to teach certain topics or which types of renewable energy to include, but the key ideas are well established. Textbooks from the 17th century are recognisable today (and, more importantly, so are the problems solved).

To identify the knowledge, use the course syllabus, textbooks and other trusted sources such as the Institute of Physics' *TalkPhysics* (2016) and *Supporting Physics Teaching* (talkphysics.org and supportingphysicsteaching.net). We can call this *core knowledge*.

But, there is other knowledge that is often missed:

- What are the archetypal questions for this topic and how do I solve them?
- What are the relevant models and how do I know when to use them?
- What are the stories behind these ideas?
- What are the common misconceptions for this topic and how can I avoid them?

In the following sections I have written about these additional types of knowledge.

Archetypal questions

Every topic in physics has its archetypal questions – the problems that are asked, in one form or another, in every physics exam. These problems are a common language for physicists – they are in every physics textbook around the world. They may be disguised using different contexts, but the deep structure and the method of solving it are the same.

The key to mastering each problem is to do it so many times in different guises that the learner can spot it without thinking about it. Recognising the archetype becomes intuitive.

Model-based problem solving

Physicists not only have heads full of problems and their solutions, they also have heads full of models and how to apply each one. For example, a physicist has several electricity models, each with different uses and limitations. She will use a simple current model for series and parallel problems, a model involving a beam of electrons for cathode ray tubes and a model of ions in a solution for electrolysis. Some of her models will be purely mathematical while others will be largely concrete. The physicist needs to know each model and the problems they can help solve.

Each of the big ideas in this book have their own models. I have described relevant ones in each chapter and shown problems they can solve.

The privileged status of stories - Willingham

Back in 2004, Daniel T Willingham wrote an article about the power of stories in our brains. (see Willingham (2004) (https://www.aft.org/periodical/american-educator/summer-2004/ ask-cognitive-scientist)) He said stories are somehow easier to understand and easier to remember and are therefore "psychologically privileged" (Willingham 2004).

Willingham identifies four Cs to help think about effective use of stories: causality, conflict, complications and character. Physics stories are full of causality, and sometimes character, but we often fail to emphasise the conflict and complications, possibly because they might distract. I try to put as much conflict and complication in as I can because that's what makes the story memorable. The history of physics is full of relevant conflict (for example, Galvani and Volta or Benjamin Franklin and the Abbé Nollet disagreeing about the nature of electricity). And there is conflict that is purely about spite (for example, Isaac Newton and Stephen Grey). If conflict is memorable (and we are in the business of developing memories), we should emphasise conflict wherever we can. And the history of physics is full of it.

Just telling the story is powerful, but it is more powerful (and accountable) to have the students write sentences about the story during or after telling it. For example:

- In this story, _____ *causes* _____.
- The *conflict* in this story is _____.
- The main *complication* in this story is _____.
- The main *character* is _____, who _____.

In each of the physics chapters in this book, I have written key stories in the development of the idea. I have chosen each story with the four Cs in mind. My aim is to use the privileged power of stories to rapidly build and develop schemata.

Misconceptions - When knowledge is wrong

Babies are born knowing physics. They express surprise when objects appear to be suspended in mid-air or pass through walls. These are the primitive physics schemata we are all born with. Onto these we add experiences from our lives: metals are cold, batteries run out of charge, the sun moves. Then in physics lessons we try to supplant this knowledge with a more formalised knowledge, often with mixed results.

All of our children come to class with heads full of unhelpful knowledge - misconceptions. These were a huge area of PhD research in the 1980s and 1990s, and as such we know a lot about them.

With the current emphasis on knowledge, the research into misconceptions becomes very relevant. One of my favourite books is *Children's Ideas in Science*, edited by Driver, Guesne and Tiberghien (you may be able to get a copy second hand). It explores the world of novice scientists' minds, rich with rational, plausible but incorrect knowledge.

More recently, Harvard's (2011) MOSART project has provided resources for teachers to identify misconceptions (it is useful for teachers to try too). You have to go through a short training process before being allowed access to the assessments. The questions are multiple choice - not for summative assessment, but to help teachers identify which of your students hold common misconceptions. Below is an example:

Scientists say a metal doorknob indoors often feels cold to you because:

1   *Cold from the doorknob goes into your hand*
2   *Heat from your hand goes into the doorknob*
3   *Cold moves from the doorknob to your hand*
4   *Heat is pulled from the doorknob by your hand*
5   *Metals are always colder than air.*

(MOSART 2011 test question)

The marking scheme tells you the percentage of students who chose the incorrect answer (A) and what the misconception is.

Recent evidence shows that our misconceptions never go away, but that we learn to select the relevant, acceptable knowledge for the situation. In other words, we all really believe that the Earth is flat and that Australia is impossible, but choose a different model in most situations. An incorrect answer may not mean the student doesn't have the knowledge - she may simply not realise she's supposed to use it for this question.

Physicists often cannot tell you how they identify the appropriate knowledge, model or technique to solve a problem. It is intuition. Nobel laureate Herbert Simon wrote about intuition: "The situation has provided a cue; this cue has given the expert access to information stored in memory, and the information provides the answer. Intuition is nothing more and nothing less than recognition" (Simon 1992). Intuition is recognition, and recognition is memory.

This is why you need to practice as many questions as possible.

Refutation texts

Misconceptions are tenacious and resilient. When you think you've got rid of one, it reappears. Long-term memories are for life. Instead of trying to remove the misconception, the solution is to recognise the misconception and build the acceptable understanding onto it. A strategy called *refutation texts* has been shown to work for this.

A refutation text is a short paragraph, written by the learner, which does three things:

1   States the misconception;
2   Explicitly says that this is not correct;
3   States the accepted scientific viewpoint.

I use sentence starters to reduce Cognitive Load, for example:

Many people believe _____.
However, _____.
Most scientists state that _____.

Using the MOSART example test question, an example refutation text is:

Many people believe that when you touch a metal doorknob, coldness moves from the metal into your hand. However, cold does not move. Scientists say that it is heat moving from your hand into the metal that makes it feel cold.

You will likely need to do this several times for each misconception using slightly different examples. It is worth spending time addressing misconceptions, because they will always come back, especially under stressful circumstances.

Practical work in physics

Science without practical is like swimming without water.

(SCORE 2008: 10)

Do you agree with this statement? Do your colleagues? Do your students? There is evidence that practical work is not an effective way of teaching content (see the Further Reading section at the end of this chapter), but it is given a high status in UK classrooms.

Using practical work is a choice – there are usually other ways of teaching whatever you are planning to teach, techniques that take less time and use fewer resources. If you can teach more efficiently using a more direct model of teaching, perhaps you should.

But if you decide to use practical activities in your lessons, you need to make it count.

Reducing Cognitive Load for practical work

Cognitive Load Theory explains why students don't learn well from practical lessons: there is too much happening at once. Learners have to: collect and assemble apparatus, follow instructions from memory or from verbal or written instructions, work collaboratively, make and record careful observations and then pack everything away. And that list does not include thinking about the science.

So how can you reduce the Cognitive Load? The simplest way to reduce Cognitive Load is for you to do the practical as a demonstration. While you are modelling the experiment, you can direct their attention to relevant details.

If you want them to do the practical work themselves, remove as much Cognitive Load as possible. What is it exactly that you want them to learn or practise? Do they need to assemble the practical themselves? Do they need to draw the results table or graph axes? Is it worth training them to get the equipment out and put it away so that it becomes automatic?

Literacy – A different sort of physics problem

What are the Cognitive Loads associated with reading and how can we reduce them?

Reading is a physics problem that doesn't receive much attention in class. I think it should. Science professionals read a lot (see Figure 0.7).

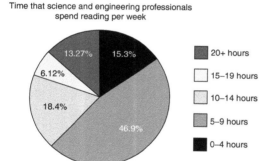

Figure 0.7 **Reading lessons for scientists**, from Ben Rogers September 2015. https://eic.rsc.org/analysis/reading-lessons-for-scientists/2010065.article

And they read to learn (see Figure 0.8).

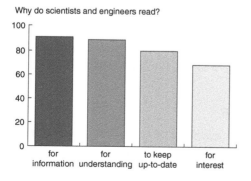

Figure 0.8 **Reading lessons for scientists**, from Ben Rogers September 2015. https://eic.rsc.org/analysis/reading-lessons-for-scientists/2010065.article

The problem is, most science, technology, engineering and maths professionals taught themselves (see Figure 0.9).

Who teaches scientists and engineers how to read professional texts?

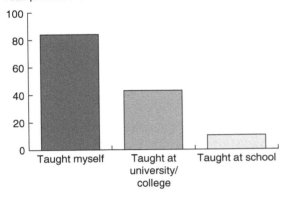

Figure 0.9 **Reading lessons for scientists**, from Ben Rogers September 2015. https://eic.rsc.org/ analysis/reading-lessons-for-scientists/2010065.article

Teaching yourself is fine, as long as there is support there when you need it. In my experience, there is little support and many fall through the net, dropping physics at GCSE or A-level because they can't access the texts. They misread exam questions, they find textbooks inaccessible and scientific papers opaque.

CLT explains why reading is difficult and suggests how to make it easier. It is difficult because all three memories are in use: long-term, working memory and external memory (the text and any scribbles added to it).

The two most important things to improve reading are in your long-term memory – or they need to be. They are vocabulary and knowledge.

Vocabulary for reading

Science teachers are excellent at teaching science vocabulary. We explain clearly, we use example sentences, we revisit, we match words to diagrams. We use every trick we know.

But we ignore key non-specialist vocabulary. Words like: *determine*, *suggest*, *establish* and *system* (I took these from a couple of recent GCSE papers). These words should be taken as seriously by science teachers as technical vocabulary.

Knowledge for reading

Along with vocabulary, the most important part of understanding is the stuff you already know: your schemata. As we read, the information in the text is held in your working memory to be presented to knowledge from your long-term memory, like a debutante or a novice speed-dater. If sense can be made, fine, but learning takes place when the long-term memory is modified, added to or contradicted.

Skills for reading comprehension

It isn't worth spending too long on generic comprehension skills. Research evidence shows that there are a few simple strategies which help, but these can be taught quickly and effectively over a few weeks. After that, you will see little improvement (see Daniel T. Willingham's (2006) article 'The Usefulness of Brief Instruction in Reading Comprehension Strategies' (http://www.aft.org/sites/default/files/periodicals/CogSci.pdf)).

In class, I focus on the following skills - or habits - that expert science readers (people like us) use most often.

To do this, I usually put students into groups of four and give them a reading card each. The student with card one reads a paragraph and then each student takes turns to finish the sentence starter on their card. Sentence starters can include:

1   *I Wonder...* Expert readers ask questions of the text. Often these questions are related to meaning, but they can be "I wonder what that word means?" or "I wonder why the writer said that".
2   *In other words...* Paraphrasing (rewording, often making clearer) is a powerful comprehension checking skill/habit.
3   *I predict...* Asking readers to predict what comes next in a text is a useful way of drawing attention to the structure and conventions of scientific texts. It is extremely useful when scanning a text for the information you want to be able to predict whether the information might be in a nearby section.
4   *So far...* Summarising is a habit which encourages prioritisation of information.

When the fourth student has finished, they each pass their card to the team member on their left and repeat the process.

If these activities can be practised enough (several times over a few weeks, with occasional top-ups) they quickly become part of a learner's reading schema - a low effort strategy to use when reading.

What are the Cognitive Loads of writing and how can we reduce them?

We can only think about two or three novel items at one time, so writing is hard. There is a lot to think about. Table 0.1 shows two things:

1   Why writing is a high-load activity.
2   How to reduce load for novice writers.

Table 0.1 Strategies for reducing Cognitive Load associated with writing

Cognitive Load	Reduce the load for novices by using...
1 Choosing the relevant knowledge and vocabulary.	Mind map/notes.
2 Planning the overall structure of the text – the argument.	Outline plans/sequencing activities (e.g. print out the individual ideas for students to sequence).
3 Planning short sequences – a couple of sentences to make a point.	Bullet points/sentence sequencing activities.
4 Structuring individual sentences.	Sentence starters/write-rewrite activities.
5 Spelling, punctuation and grammar.	Teach model sentences/remind students how to punctuate / practise punctuating sentences correctly.

All of the Cognitive Load activities are important; practice them one at a time. Decide which part of writing you want your students to develop and reduce the Cognitive Load from the other elements.

The following writing task is an analysis of the Cognitive Load imposed by a typical physics writing task.

> **Writing task:** Explain why a skydiver reaches terminal velocity.

What is the knowledge required by this question?

1   Key knowledge:

    a   Gravity

    b   Air resistance

    c   Velocity increases

    d   Resultant force decreases

    e   Balanced forces

    f   Terminal velocity.

2   Outline:

    a   At first – no motion

    b   Speed increases

    c   Until terminal velocity.

3   Sequence of sentences:

    a $v = 0$, acceleration due to gravity. No air resistance.

    b v increases $\rightarrow$ air resistance increases $\rightarrow$ resultant force decreases $\rightarrow$ acceleration decreases.

    c $v =$ forces balance $\rightarrow$ acceleration $= 0 \rightarrow$ terminal velocity.

4   Key sentences:

   a   Initially, the velocity is zero, so the air resistance is zero and the skydiver accelerates.
   b   As the velocity increases so does the air resistance, resulting in a decreased acceleration.
   c   When the air resistance balances the force due to gravity, the acceleration reaches zero – this is the skydiver's terminal velocity.

5   Spelling and grammar (SpaG) check.
6   Proofread and edit.

How to teach writing in physics

Writing is one of the best ways to ensure your students are thinking, and to get an insight into their thoughts. But because writing has a very high Cognitive Load, it is worth separating each element of writing out: work on one element at a time and assess one element at a time. These elements include: identifying relevant concepts, arranging these concepts in a logical sequence, constructing well-formed sentences and structuring the writing. CLT has revealed several useful techniques to make learning through writing more effective:

1   Start with model answers. Greg Ashman (2017) recommends modelling followed by near identical problems for students to complete (gregashman.wordpress.com/2017/05/13/four-ways-cognitive-load-has-changed-my-teaching/).
2   Use completion problems. Choose which type of Cognitive Load in Table 0.1 (see page 17) you want your students to focus on and reduce the rest.
3   Use gap fills, sentence starters and mind maps as effective techniques to reduce Cognitive Load. Later you will want to demonstrate combining the different elements into a coherent piece of writing.
4   Write-rewrite. This *Reading Reconsidered* technique (Lemov *et. al.* 2016) is especially effective in reducing Cognitive Load when writing complex scientific sentences. Students write their sentence once, share effective answers and then rewrite. Your students are effectively unloading Cognitive Load as the first draft, allowing greater intellectual resources to be applied to the second draft.
5   Reduce support. As the student develops expertise, these strategies eventually increase Cognitive Load as they get in the way.

Conclusion

Learning physics means learning to solve the problems of physics. Cognitive Load Theory provides a model and strategies to make learning to solve the problems of physics more effective.

   The five remaining chapters apply Cognitive Load Theory to five big ideas of physics: electricity, forces at a distance, energy, particles and the universe, always starting with the stories.

Further reading - A big idea about learning

- *Bringing Words to Life*, Beck, McKeown and Kukan, The Guilford Press. 2002
- *Children's Ideas in Science*, Driver, Guesne and Tiberghien, Open University Press, 1985.
- Cognitive Load Theory: Research that Teachers Really Need to Understand, Centre for Education Statistics and Evaluation, www.cese.nsw.gov.au/images/stories/PDF/cognitive_load_theory_report_AA1.pdf.
- *Feynman's Tips on Physics*, Basic Books, 2013.
- MOSART (Misconceptions-Oriented Standards-Based Assessment Resources for Teachers) www.cfa.harvard.edu/smgphp/mosart/
- *Practical Work for Learning*, The Nuffield Foundation, www.nuffieldfoundation.org/practical-work-learning.
- Practical Work in Science: Misunderstood and Badly Used? Jonathan Osborne, *SSR*, September 2015.
- Practical Work: Making It More Effective, Robin Millar and Ian Abrahams, *SSR*, September 2009, www.gettingpractical.org.uk/documents/RobinSSR.pdf.
- Reading Lessons for Scientists, Ben Rogers, September 2015, https://eic.rsc.org/analysis/reading-lessons-for-scientists/2010065.article.
- *Reading Reconsidered*, Lemov, Driggs and Woolway, Jossey-Bass, 2016.
- The Association for Science Education (ASE) published two editions of their journal *School Science Review (SSR)* on practical work in June and September 2015. ASE members can download the papers from the website at https://www.ase.org.uk/journals/school-science-review/.
- The Learning Scientists, Six Strategies for Effective Learning: Materials for Teachers and Students, www.learningscientists.org/downloadable-materials/.
- *The Reading Mind*, Willingham, Jossey-Bass, 2017.
- *The Writing Revolution*, Hochman and Wexler, Jossey-Bass, 2017.
- *Thinking, Fast and Slow*, Kahneman, Penguin, 2011.
- Why Minimal Guidance During Instruction Does Not Work: An Analysis of the Failure of Constructivist, Discovery, Problem-Based, Experiential, and Inquiry-Based Teaching.' Kirschner, Sweller and Clark. www.cogtech.usc.edu/publications/kirschner_Sweller_Clark.pdf.

1 Electricity

Electricity: The versorium needle, Guericke's Sulphur Sphere, Gray's Dangling Boy, The Leyden Jar, Galvani's animal electricity, Volta's Electric Pile, Ørsted's Needle and Faraday's Motor.

Introduction

Electricity is learnt, like all abstract concepts, through a gradual accumulation of experience and knowledge. We seldom learn about electricity through words: we understand electricity through diagrams, problem solving, models and practical work with circuits.

So, in most physics classrooms, we neglect what Daniel Willingham refers to as "the privileged status of story" (Willingham 2004). Without the stories, electricity can become dry and functional – there's little spark. The abstract needs something human to adhere to.

That is why a narrative is so important. In this chapter, I have selected eight inventions to tell the story of electricity from 1600 until 1839. After 1839, the narrative of electricity merges with the story of particles, so the story continues in Chapter Four.

Following the story section, this chapter moves into the classroom. Electricity suffers from very specific terminology, pretending to be everyday words: charge, current and electricity. Although I have tried to explain these terms clearly, explanations are not the answer: the answer is exemplars – the standard problems that every physicist has solved.

Students of electricity also have to overcome persistent misconceptions – electricity seems to have more than most. I have listed a few, and suggested strategies for overcoming them.

An exploration of four commonly used models follows. I think we should teach all of them and support learners to compare and critique them all. With knowledge, more is more.

Finally, I have finished the chapter by describing key classroom practicals: why it is useful, what can go wrong and how to fix problems.

A history of electricity

The versorium needle - 1600

These are the utterly false and disgraceful tales of the writers.

(Gilbert 1600: 48)

This history begins, like all science, with an error: "Amber, when rubbed, will not attract dried basil" (Alexander Aphrodiseus, cited by Gilbert 1600: 48). It isn't true and it is very simple to prove – just rub some amber on cloth and watch it attract dried basil. You can try it yourself. But the error persisted for more than 2,000 years, until it was disproved in 1600 by William Gilbert of Colchester.

In 1600, William Gilbert began the modern study of electricity, publishing a book which presented a true account of electricity and magnetism: *De Magnete*. The quote at the top of this page is from this book. Errors like the amber/basil mistake clearly made Gilbert cross.

To investigate this phenomenon, Gilbert invented a mysterious instrument: the versorium needle, as shown in Figure 1.1.

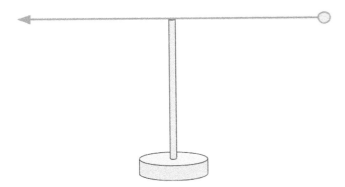

Figure 1.1 The Versorium needle

The needle is made from un-magnetised metal. When rubbed amber is brought towards the versorium, the needle will turn until it points towards the amber. Gilbert could not explain how the versorium worked (Stephen Gray demonstrated electrostatic induction with his 'dangling boy' 130 years later), but he used it to prove more Ancient Greek statements wrong.

The Greeks believed that amber alone had the property of attracting when rubbed. Gilbert used his versorium to prove that many other materials showed the same property. His list of materials included: diamond, sapphire, glass, sulphur, sealing wax and resin. He named these materials electrics because they behave like amber ('elektron').

So Gilbert resolved the great amber/basil error and demonstrated that many more substances can be charged than amber.

Gilbert (1600: 113) also contributed to the great chain of scientific errors by introducing one of his own:

All electricks attract all things: they never repel or propel anything at all.

Gilbert's great mistake was not recognised until 1663, when a remarkable scientific inventor, Otto von Guericke, demonstrated repulsion using his electrical sulphur sphere.

Museum of electrical history

A versorium needle is very simple to make. The set-up is shown in Figure 1.2.

1   Cut an elongated diamond of aluminium foil (about 75mm long and 30mm wide) and fold along the long line of symmetry. This is the needle.
2   Balance the needle on a cocktail stick set in a cork or clay.
3   Test various electrics (insulators) after rubbing with cloth. The needle should react clearly.

Figure 1.2 Gilbert's versorium needle

How the versorium needle works

At the end of the 16th century, William Gilbert invented a device for detecting electric charge. He called it the versorium needle. When a charged object is brought towards the needle, the needle will point to the charge.

We understand this now, because we know that charges in the metal (electrons) can move. When a positive charge is brought near the needle, the electrons are attracted towards it, making the needle's point negative. The negative point is attracted towards the charged object.

If a negative object is brought near the needle, the electrons move to the far end of the needle, leaving the point with a positive charge. The positive point is attracted towards the charged object.

But Gilbert didn't know that.

Guericke's sulphur sphere - 1672

It seems reasonable to suppose that if the Earth has a fitting and appropriate attractive potency it will also have a potency of repelling things that might be dangerous or disagreeable to it.

(von Guericke 1672: chapter 6)

Otto von Guericke was extraordinary. He is most famous for inventing the air pump and using it to demonstrate the enormous effect of atmospheric pressure with his dramatic Magdeburg Hemispheres (proving, finally, that *horror vacui* or *'nature abhors a vacuum'* is not true). He also invented the electrical generator.

In 1663, Guericke made a hollow ball of sulphur that could be rotated. When the operator placed a dry hand (Guericke had famously dry hands) onto the rotating sphere, the sphere became charged. Dry human skin is especially effective, though other materials, including wool and leather, also work well.

The apparatus was made by blowing a glass sphere ("about the size of a child's head" (von Guericke 2012: 227) and using it as a mould for molten sulphur. When the sulphur solidified, the glass was broken, producing a hollow sulphur ball.

Guericke noticed that the sphere first attracted chaff, but once the chaff touched the sphere, it would then be repelled. He noticed that a feather was first attracted to the sphere, and then, once contact had been made, the feather was repelled back to the ground from where it was again attracted. A feather could make the journey repeatedly.

Even though Guericke's sulphur sphere was unreliable to use and expensive to make, electrical philosophers across Europe rushed to build their own. One experimenter stopped at the glass mould stage and tested the glass sphere instead. It produced charge even more effectively than the sulphur.

Some say Isaac Newton was the inventor of the glass globe generator. It was a small development, and if Newton discovered it, it was the only useful thing he did for electricity. His behaviour towards other electricians was jealous, spiteful and obstructive, especially towards Stephen Gray, the hero of the next section, with his 'dangling boy'.

Museum of electrical history

A glass generator can be made using a clean glass jar, as per Figure 1.3.

1   Glue an axle onto the top and bottom of the jar (I used lego).
2   Mount each end of the axles in a frame (again, I used lego).
3   Put a conducting brush (aluminium foil) onto the jar connected with a conducting clip.
4   Attach a bundle of threads to the top of the clip to show charge.
5   Rotate the jar and press an insulating material onto the jar at the same time. Guericke used his dry hands, but my hands seemed to be too well moisturised. I found wool worked better.

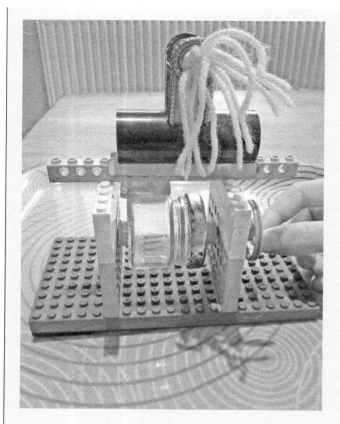

Figure 1.3 A kitchen recreation of Guericke's electrostatic generator (see https://twitter.com/benrogersedu/status/766298265608450048)

How Guericke's sphere works

In 1663, Guericke made the first electrical generator. When two insulating materials touch, one of them will attract electrons more strongly than the other, causing one to gain a negative charge and the other a positive charge. When you rub two insulators together, the charge builds up.

But Guericke didn't know that.

Gray's 'dangling boy' – 1730

Figure 1.4 shows a boy suspended by silk cords, charged by a Guericke sphere. The boy's body attracted feathers and leaves of brass. The boy was a pupil at Charterhouse School. The experimenter was Stephen Gray, the discoverer of electrical conduction.

Figure 1.4 Gray's dangling boy demonstration (from a German publication in 1744 describing Gray's work)

Stephen Gray had become a victim of Newton's domineering behaviour following an argument over data. The great seventeenth century astronomer, John Flamsteed, was producing the most extensive and accurate star catalogue ever made. Newton used and published Flamsteed's data without permission. Newton was unapologetic. Flamsteed was furious:

> I had resolved aforehand his knavish talk should not move me. I complained then of my catalogue being printed by Raymer, without my knowledge, and that I was robbed of the fruit of my labours. At this he fired, and called me all the ill names, puppy, etc., that he could think of.
>
> (Maunder 1990: 54)

Newton was not a man to build bridges. This was unfortunate for Stephen Gray. Flamsteed was Gray's great supporter and any friend of Flamsteed was an enemy of Newton. Newton stopped Gray's career dead.

Despite great early successes and influential friends, Gray's great work had to wait until Newton died before resuming. In his sixties and living as a pensioner in Charterhouse, a home for destitute gentlemen, Gray returned to his electrical studies. He began experimenting with a glass electrical generator. The glass cylinder was sealed at one end with a cork to keep the dust out. Gray was "much surprised" when he held a feather to the cork and discovered that the feather was attracted and repelled many times – just as it was by the rubbed glass. He concluded that an "attractive virtue" was "communicated to the cork by the excited tube" (Gray 1731: 19).

He then investigated how far the electrical 'virtue' could be transmitted through wood, as demonstrated by Figure 1.5.

> Having by me an Ivory Ball of about one Inch three Tenths Diameter, with a Hole through it, this I fixed upon a Fir-Stick about four Inches long, thrusting the other End into the Cork, and upon rubbing the tube, found that the Ball attracted and repelled the Feather with more Vigour than the Cork had done.
>
> (Gray 1731: 10)

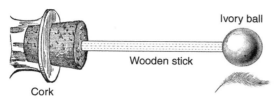

Figure 1.5 Gray's discovery of electrical conduction

Gray realised that electricity was not confined to the glass; it was able to travel along certain materials. He began investigating which materials could carry the 'electrical virtue':

> Then I made use of first iron, and then brass wire, to fix the ball on, inserting the other end of the wire in the cork, as before, and found that the attraction was the same as when the fir-sticks were made use of, and that when the feather was held over against any part of the wire, it was attracted by it, but though it was then nearer the tube, yet its attraction was not so strong as that of the ball.
>
> (Gray, 1731: 21)

Gray had discovered electrical conductors and insulators.

He extended his investigations further. Dangling eight metres of thread from a balcony to the floor below, he connected the top end to the generator and the bottom to the ivory ball. The ball attracted feathers. The string conducted the electricity.

Gray wanted to experiment with even longer threads, but did not have a high enough balcony. Instead he decided to lay the thread horizontally along a wooden floor. This did not work.

Eventually, he and Reverend Wheeler (a friend and supporter) suspended the conducting thread on fine dry silk threads. Using the silk as insulation, the range they were able to carry the electricity increased to more than one hundred metres.

Electrical demonstrations ('electrickery') were popular in Enlightenment England. Stephen Gray published his results, and, I hope, made a little money, with an astonishing display of his electrical discoveries.

Charterhouse, the home for retired impoverished gentlemen, had a school. This was convenient. Gray created an electrical demonstration that became famous across Europe and the new world. This demonstration involved dangling a small boy on silk strings, then connecting him to a generator. Charterhouse School provided the (presumably willing) volunteers.

Anyone who has had the opportunity to charge children to thousands of volts will appreciate the skill and showmanship of the 'dangling boy' demonstration.

The boy was suspended from a wooden frame on the insulating silk strings and connected to the rotating glass generator by his foot (refer back to Figure 1.4). Feathers and chaff were attracted to the boy's face from the ground. To anyone interested in electrical philosophy, this was amazing: electricity could be conducted.

Luckily for the boy, the full power of electricity to cause shock and pain had yet to be discovered. It took another fifteen years for the deadly Leyden jar to be invented, and then it was invented twice.

The Leyden jar - 1745

The Leyden jar - storing electricity and providing philosophers with a clue about the nature of electricity.

A spinning glass cylinder can generate thousands of volts, but very little energy - just enough to cause little crackling sparks. Imagine the surprise if you accidentally found a way to store large quantities of electricity in a jar (and the greater surprise when the electricity discharged through you).

> I wish to inform you of a new but terrible experiment, which I advise you on no account personally to attempt. Suddenly I received in my right hand a shock of such violence that my whole body was shaken as by a lightning bolt… the arm and body were affected in a manner more terrible than I can express. In a word, I believed I was done for.
>
> (van Musschenbroek 1746: 17)

Naturally, everyone wanted a go.

The electrical collecting jar was discovered accidentally, twice in the same year, first by Ewald Georg von Kleist of Farther Pomerania and then by Peter van Musschenbroek of Leiden (using a beer glass). It is a dangerous thing and a miracle that no one was killed. It became a phenomenon.

> the first time he tried the Leyden experiment, he found great convulsions by it in his body; and that it put his blood into great agitation; so that he was afraid of an ardent fever, and was obliged to use refrigerating medicines. He also felt a heaviness in his head, as if a stone lay upon it. Twice, he says, it gave him a bleeding at his nose, to which he was not inclined; and that his wife (whose curiosity, it seems, was stronger than her fears) received the shock only twice, and found herself so weak, that she could hardly walk.
>
> (Priestley 1767: 83)

The Leyden jar was charged with a Guericke sphere. The charge was carried to a pin entering the water of the jar. The outer conductor was your hand. As the sphere turned, the charge slowly built up in the jar. The shocking moment of discovery came when you touched the pin and all of the electricity discharged. The set-up is demonstrated in Figure 1.6.

The Leyden jar became the heart of the philosophical question: what is electricity?

At the time, philosophers believed that the electricity was stored in the jar like water, but the ability of electricity to attract and repel was a mystery.

In France, an answer was proposed by Charles François de Cisternay du Fay. Du Fay and his colleagues were convinced that there were two types of electricity: vitreous (obtained by rubbing glass) and resinous (by rubbing resin). Resinous electricity attracts vitreous. Vitreous repels vitreous and resinous repels resinous.

Du Fay explained how the Leyden jar worked using his idea of the two types of electricity. To give the jar a vitreous charge, the vitreous electricity travels down the chain into the water. Resinous charge in the hand is attracted to the vitreous charge in the water. The resinous charge in the hand attracts more resinous charge down the chain. The jar now contains far more charge than has been possible before.

Figure 1.6 The Leyden jar (from Augustin Privat Deschanel (1876) *Elementary Treatise on Natural Philosophy, Part 3: Electricity and Magnetism*, D. New York: Appleton and Co., translated and edited by J. D. Everett, p. 570, figure 382)

British philosophers, on the other hand, took sides with an extraordinary gentleman from the colonies, a printer, politician and philosopher: Benjamin Franklin. Franklin preferred simplicity. He proposed a single type of electricity. When an object has too much electricity, it is positive; too little and it is negative.

Franklin explained the properties of the Leyden jar with his single type of electricity. To give the jar a positive charge, positive electricity travels down the chain into the water. The positive electricity in the hand is repelled, leaving the hand negative. The negative hand attracts more of the positive electricity down the chain. The jar now contains far more electricity than has been possible before.

> So wonderfully are these two states of Electricity, the plus and the minus, combined and balanced in this miraculous bottle!
>
> (Franklin 1747: 181)

Franklin enjoyed the Leyden jar enormously. He invented a game called "treason," which involved an electrified portrait of the king with a removable gilt crown. Anyone who tried to remove the crown while holding the gilt edge of the picture would be shocked.

His electrical fun nearly ruined one Christmas when he used Leyden jars to kill the turkey for the entertainment of his friends:

> I have lately made an experiment in electricity that I desire never to repeat. Two nights ago, being about to kill a turkey by the shock from two large glass jars, containing as much electrical fire as forty common phials, I inadvertently took the whole through my

own arms and body, by receiving the fire from the united top wires with one hand, while the other held a chain connected with the outsides of both jars.

(Franklin 1840: 255)

This lesson might have stopped other philosophers from persisting with such dangerous experiments, but not Benjamin Franklin. In 1752, he carried out the most famous of dangerous experiments: flying a kite in a thunderstorm. He connected the kite string to a Leyden jar to investigate whether the lightning could be stored as electricity. It worked. It is astonishing that Franklin survived long enough to die in his own bed, aged 84.

The Leyden jar did not have any more great secrets to reveal. It took another four decades for the philosophy of electricity to make a significant leap forwards. This time with frogs' legs and animal electricity.

Museum of electrical history

To create your own mini-Leyden jar, as shown in Figure 1.7: Warning: A Leyden jar is a capacitor. Although it's capacitance is very small, if you charge it with static electricity at high voltage, the energy stored can be enough to cause a painful shock. DO NOT CHARGE with any type of static electricity generator (e.g. a Van de Graaff).

Figure 1.7 A kitchen demonstration of the Leyden jar (do not charge this – it is for demonstration only

1   Get a small plastic jar with a lid (the less surface area, the less capacitance, result-
     ing in less charge stored).
2   Drill (or melt with a hot pin) a small hole in the top.
3   Insert a wire through the hole.
4   Fill the jar with water and a little salt.

How it would be charged (if anyone were foolish enough to try):

5   While holding the jar in your hand (your hand is the earthed plate in the capacitor),
     touch the exposed wire to the glass generator (*do not use a Van de Graaff!*).

A modern explanation

The Leyden jar is a capacitor. It stores charge according to the formula: $Q = CV$, where
Q = charge, C = capacitance and V = applied voltage.

The capacitance of an 18th century Leyden jar is approximately 1nF (1×10^{-9} Farads).
A rotating glass sphere can generate thousands of volts (a Van de Graaff generator can
reach 100,000V, - *so do not use one with a Leyden jar!*).

The formula shows that the Leyden jar will typically store microcoulombs - enough
to cause a nasty shock.

Galvani's frogs' legs and animal electricity - 1791

Galvani's animal electricity

Anyone who has experienced an electric shock (by 1791, that was pretty much anyone who
was interested) knows that electricity has a strong effect on muscles. Many philosophers
were studying the effect. Luigi Galvani's laboratory was a Doctor Frankenstein's workshop of
dissected frogs, Leyden jars and electrical generators, as per Figure 1.8.

One charming story claims that Galvani's wife was preparing frog leg soup when her knife
touched the nerve that connects a frog's leg muscle to its spine. The leg muscle twitched.
I assume that Italian housewives had noticed this phenomena for generations. Luckily, this
time, the attentive and observant Luigi noticed the twitch and the rest is history. Galvani's
version is different:

I dissected a frog and prepared it. When one of my assistants, by chance, lightly applied
the point of a scalpel to the inner femoral nerves of the frog, suddenly all the muscles
of the limbs were seen so to contract that they appeared to have fallen into violent tonic
convulsions.

(Galvani 1953: 24)

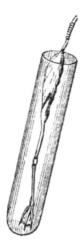

Figure 1.8 Galvani's frog leg (from Jabez Hogg (1861) *Elements of Experimental and Natural Philosophy*, figure 348, London: Henry G. Bohn)

This was the beginning of many experiments involving brass hooks, steel blades, lightning, electrical sparks and many, many frogs' legs.

> Excited by the novelty of the phenomenon, we began to make tests and experiments of various kinds, but always using the same scalpel. Nor was this additional diligence without its reward, for we discovered that the answer to the problem lay in the part of the scalpel we held in our fingers. Since the scalpel had a bone handle, we found that when this handle was held in the hand, no movements were produced at the discharge of a spark. They did occur, however, when the fingers touched the metal blade or the iron nails that secured the blade of the instrument.
>
> (Galvani 1953: 25)

The conclusion reached by Galvani was that muscles were like small Leyden jars. He named the electricity stored in the muscles 'animal electricity'. Muscles could be made to twitch by either touching the nerves with conductors or by applying electricity directly from external Leyden jars.

This was controversial from the start.

Volta's electric pile - 1799

In which Allessandro Volta proves that Galvani's 'animal electricity' can be made mechanically using his pile.

Galvani's greatest and politest rival was Alessandro Volta. Initially Volta found Galvani's explanation convincing, however his own experiments convinced him that the 'animal electricity' hypothesis was wrong. Volta believed that the electricity did not come from the muscle and that the frog's leg was responding to electricity generated somehow outside the frog.

Volta noticed that the leg twitched when two dissimilar metals (for example, copper and iron) were attached to the nerve and muscle. He realised that it was the two different metals that somehow caused the electricity to flow.

This led Volta to much experimenting with arrangements of different metals, until he developed his famous pile. The pile is a stack of disks: copper, zinc and card soaked in brine. This pattern is repeated. His invention proved that electricity could be made by chemical means and required no 'animal electricity'.

The Voltaic pile, like Guericke's sulphur globe and the Leyden jar, opened up many fresh opportunities for scientific philosophers to explore, including the decomposition of water into hydrogen and oxygen; the discovery of the chemical elements sodium (1807), potassium (1807), calcium (1808), boron (1808), barium (1808), strontium (1808) and magnesium (1808) by Humphry Davy and also the next major discoveries in electricity by Oersted and Faraday.

Museum of electrical history

It is simple to make a voltaic pile, as shown in Figure 1.9. You will need aluminium foil, copper foil (or shiny copper coins), filter paper or card and saturated salty water with a few drops of vinegar (or other acid). You can test the pile with a voltmeter or an LED.

Figure 1.9 A kitchen recreation of a voltaic pile

1   Make your disks (at least five of each - the more disks, the higher the voltage). They should all be the same diameter. Larger disks provide larger currents.
2   Soak the paper/card disks in the salty acidified water.
3   Stack the disks in the order of aluminium, paper, copper, aluminium, paper, copper etc., until all the disks are used.

To test your pile using an LED - touch one leg to the top and one to the bottom. When you get the legs the right way around, the LED should light up (possibly dimly).

Oersted's twitching needle - 1820

Scientific discoveries are usually private affairs. I imagine Stephen Gray muttering "That's odd...". I suspect Musschenbroek made a bit more noise when he accidentally discovered the Leyden jar, but we don't know because he was alone. Oersted made his great electrical discovery in a crowded lecture room full of people who wanted to see the latest electrical discoveries. But no one in the audience noticed.

Oersted was delivering a lecture in Copenhagen, setting up the apparatus for a demonstration. He was a popular lecturer and the theatre was full. When he connected his voltaic pile to complete a circuit, he noticed that the needle of a compass on the table moved, as in Figure 1.10.

Philosophers had been trying to get this effect for years, but they had been doing it wrong. Common sense says that if an electrical current can make a magnetic field, the force should make the compass line up with the wire. Oersted's needle twitched the wrong way.

It was only a small twitch. No one in the audience saw it. Hundreds of people were in the room and only Oersted knew what had just happened and he didn't make a big fuss about it either. Oersted knew what he had seen: electricity could make magnetism.

And magnetism can make things move. But that would be Michael Faraday's electrical invention 19 years later.

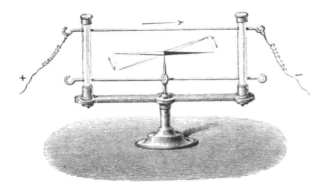

Figure 1.10 Oersted's discovery (Wikimedia commons)

Museum of electrical history

Oersted's experiment is very simple to recreate, as shown in Figure 1.11. A simple series circuit with a switch will cause a compass needle to deflect.

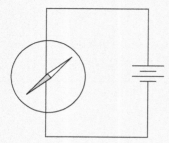

Figure 1.11 A diagrammatic representation of Oersted's apparatus

Faraday's motor - 1839

> When we consider the magnitude and extent of his discoveries and their influence on the progress of science and of industry, there is no honour too great to pay to the memory of Faraday, one of the greatest scientific discoverers of all time.
>
> (Rutherford 1931, cited by Rao 2009: 281)

In his lifetime, Faraday refused all honours. He declined a knighthood and refused to become President of the Royal Society. But in his early years, Faraday was not considered a gentleman. When travelling through Europe with Humphry Davy and Davy's wife, Faraday had to travel on the outside of the coach.

It was Davy, though, who gave Faraday his early opportunities. Faraday was a book-binder's apprentice when Davy offered him a post as his assistant at the Royal Institution. Soon afterwards, Davy arranged for Faraday to become Chemical Assistant at the Royal Institution.

Faraday's early work was chemical (and significant), but he is best known for his work on electricity and magnetism.

The relationship between Faraday and Davy survived the uncomfortable carriage trip around Europe (and Mrs Davy's refusal to eat with him), but did not survive Faraday's success with the electric motor (Faraday forgot to credit Davy's earlier unsuccessful attempt at making a motor). The controversy resulted in Faraday's assignment to other projects.

But Faraday was not finished with electricity and magnetism. He was the first to reverse Oersted's process, producing electricity by moving a conductor in a magnetic field.

So Faraday created the first motor and the first electrical generator. This is what Rutherford was referring to when he praised Faraday's contribution to industry. Faraday's contribution to physics though will be continued in Chapter Two: Forces at a distance, as it was Faraday who proposed the lines of magnetic field.

The story of electricity continues in different chapters. The discovery of the electron belongs in Chapter Four: *Particles*. Faraday's contributions continue in the chapter: *Forces Acting at a Distance*. The story of physics is a complicated soap opera.

Museum of electrical history

A Faraday motor ('homopolar') is a lovely thing to make, as demonstrated in Figure 1.12. You need some small Neodymium Magnets, a battery cell (AA or AAA) and some copper wire (not insulated).

1   Sit the cell on top of two or three of the magnets.
2   Make a wire spiral with the top bent in to sit on the top of the cell and a small loop to fit loosely around the magnets at the bottom.
3   When current flows through the wire, it should rotate. It may take some adjustment to make sure the wire is free to rotate without too much friction while still maintaining electrical contact with the top of the cell and the magnets at the bottom.

Warning: The wire can get hot as these motors are very inefficient.
Try experimenting with other shapes for the wire.

Figure 1.12 A kitchen version of Faraday's motor

Electricity in the classroom

Key concepts

Electricity has three concepts which hold the key to understanding the subject: voltage, current and charge. These cannot be taught one-by-one; an effective understanding is built up by repeated exposures to these three ideas over years.

But I will begin by examining the word 'electricity' itself.

Electricity's colourful past

Among physicists, the word 'electricity' has lost its usefulness. It hangs about, rather a nuisance: an idea with a colourful past. But that past is glorious.

The word 'electrical' was first used by Thomas Browne in Norwich to mean insulators which can be charged by rubbing (like amber – the word 'electrum' is Latin for amber). Here is Browne's list of electrical bodies (which he begins with Gilbert's list, presumably tested using his versorium 40 years earlier):

> Now, although in this rank but two were commonly mentioned by the ancients [amber and jet], Gilbertus discovereth many more; as diamonds, sapphires, carbuncles, iris, opals, amethysts, beryl, crystal, Bristol stones, sulphur, mastic, hard wax, hard resin, arsenic, sal-gemma, roche alum, common glass, stibium, or glass of antimony. Unto these, Cabeus addeth white wax, gum elemi, gum guaiaci, pix hispanica, and gypsum. And unto these we add gum animi, benjamin, talcum, china-dishes, sandaraca, turpentine, styrax liquida, and caranna dried into a hard consistence.
>
> (Browne 1650: 326)

Browne was thorough. He tested each substance himself. I have found it much harder to obtain these substances in Norwich today.

While philosophers developed the concept of electricity, showmen quickly realised the potential for drama: the shocks, the sparks, the 'electrickery'. Many of the philosophers adopted the showmen's circus style, entertaining friends and society with electric kisses, dangling boys and electrocuting turkeys for thanksgiving.

Then came Galvani's twitching frogs' legs and animal electricity. Galvani's nephew toured Europe, making the recently executed twitch and open their eyes. The idea that electricity was linked to life was used by radicals in the early 19th century to sermonise against God and the natural order.

Then there were the quacks and medical charlatans with their electropathic belts and Pulvermacher's Galvanic bath chains.

So, it is a relief that the word electricity has fallen into disuse in modern physics, fit only as the title for courses. It is not a word to be used carelessly.

The next three words – charge, current and potential difference – benefit from clear explanations, but explanations are not how these concepts are learnt. They are only really understood by answering as many 'end-of-the-chapter' questions as possible. It is the exemplars we have learnt that make us physicists.

Electric charge

Charge is the most fundamental, and the most slippery, of the electrical terms. It is defined as follows: "Electric charge is the physical property of matter that causes it to experience a force when placed in an electromagnetic field" (Wikipedia 2017).

The definition alone is hopeless.

In physics, the word *charge* (Figure 1.13) is used in at least three different ways. First, and most important, *charge* is used as an uncountable noun - physicists often talk about charge as though it were an amount of a substance, like water or plasticine, not separated into separate droplets or chunks.

This idea developed because physicists initially understood electricity as a flow of charge, imagining water flowing in a pipe. This uncountable use of charge is important, because it allows us to use and understand current, voltage, electric fields and the conservation of charge.

Another useful, important and tricky concept that uses this version of the word charge, is *unit charge*. A unit charge is a quantity of charge that we give a value of one (analogous to a unit mass - the amount of material that will give us a mass of one).

The concept of discrete *charged particles*, such as electrons, protons and ions, developed later, introducing the next usage of the word charge: this time, charge means *charged particle*. We talk about a flow of charges, meaning a flow of electrons or ions. This idea introduces the confusing idea that current flows in the opposite direction to the flow of electrons in a circuit. The way out of the confusion is to remember that charge and charges are not the same thing.

The third use of the word charge is as a verb: *to charge*. It can be used correctly to mean 'to put charge on to an object', for example when a balloon is rubbed on a jumper, or incorrectly, such as charging a battery (when we really mean putting energy into a battery - this is the word *charge* we all use every day!).

We also use *charged* as an adjective (for example charged particle, charged capacitor, charged balloon and, unfortunately, charged battery).

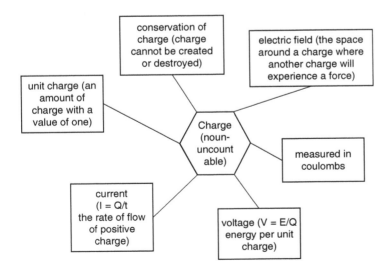

Figure 1.13 An analysis of the word 'charge'

You can see how the definition of charge is unhelpful. If you need to teach the definition, just make sure they learn the exam version as found in the syllabus specification. Otherwise, help your students construct a rich understanding, through the use of a mixture of models, problems to solve, questions to answer and practical work. Be explicit - charge is a challenging concept that many of your students won't develop on their own or by chance - and be very patient.

Current

Current is a nice simple word: it means *the rate of flow of charge*. So you need to explain *rate* to get a proper understanding. And there are two parts to rate: the velocity of the charge and the amount of charge per unit volume. So then you need to explain what the *amount of charge per unit volume* is.

This is where models come in. And I recommend laying the models on thick for current. The classic model is the flowing water model, where water is the charge, flowing through pipes. You can demonstrate the 'rate of flow' by filling a beaker with a narrow jet of fast-flowing water or a wide jet of slower water.

The Institute of Physics favours the rope model, where the rope is the charge, travelling round the circuit. You can easily increase the current by increasing the speed, but it isn't so simple to increase the charge per unit volume (idea for thought - use fatter rope?). The advantage of this model is the idea that current doesn't get used up or lost.

A third model commonly used is the human model of charge moving round the circuit. Move the desks to make a circuit. Get one student to wear a light bulb hat and another to wear a cell hat. The rest of the students are charges, moving round the circuit, with the light bulb lighting up and the cell pushing the students gently round.

Each of these models has strengths and weaknesses, but isn't that the point? By comparing the models, students develop a critical understanding of the properties of electricity denied by the use of a single model. Be explicit that the models don't all illustrate the same thing. Don't oversimplify for the sake of understanding, or if you do, don't do it for long. If you don't reintroduce the complexity, their understanding will be impoverished.

Potential difference

Potential difference comes with many names: voltage, emf, tension. They all mean *the amount of energy carried by a unit charge* (an abstract quantity carried by a different abstract quantity). To get a sense of potential difference, students need to experience the effects of voltage (you could use a Van de Graaff generator - many thousands of volts - and simple circuits). Models can help. Finally, students need to practice answering as many voltage questions as possible: the concept of voltage accumulates with practice.

Misconceptions

Because the terminology of electricity is used, and misused, commonly in everyday life, unpicking misconceptions is more challenging in electricity than any other topic. The classic text I recommend is *Children's Ideas in Science*, edited by Driver, Guesne and Tiberghien (1985) – every science department should have one. I have outlined three of the most common misconceptions.

Something gets used up going around the circuit, but what?

As charge travels around a circuit, it loses energy (though it is better to say that energy is transferred rather than used up). The potential difference does get smaller as the charge travels around, because the charge is losing energy. However, neither charge nor current decreases (or gets used up) – these both stay the same. This is because charge is made of particles (usually electrons) and cannot be destroyed.

What do batteries and cells have that gets used up?

Because we talk about charging batteries, it is common for people to believe we are pumping extra charges into the battery. This isn't true. When a battery is 'charged' we really mean that the charge has been given energy. In other words, the electrons have been moved from positions in the cell with low energy to positions with high energy – the same number of electrons, but greater energy. So the energy of a battery gets 'used up' (or transferred), but the charges don't.

How does charge give up its energy when travelling around a series circuit?

Most learners predict, upon seeing Figure 1.14 that the charge will *dump* its energy at the first bulb of a series circuit. This doesn't happen. Energy is transferred at each bulb, depending on the resistance of each bulb. The simplest way to explain this is using the rope model (see next section: Models).

What is the best way to overcome misconceptions?

You cannot expect to overcome a misconception with one brilliant lesson. Misconceptions are too deeply rooted in long-term memory for that. Instead, you need to use repetition and practice to build strongly embedded and correct knowledge.

Weekly quizzes are quick and effective. Practice questions work well, but make them too easy rather than too hard – you want them to learn the right answers, not the wrong ones. Strong memories are also made when learners write refutation texts, for example:

> Many people believe that a cell or battery holds all of the charges inside and lets them out through the circuit. This isn't true. The cell is more like a pump, pushing the charges round the circuit.

> The cell or battery doesn't run out of charge, it runs out of energy. We should probably say that we are energising a battery, instead of saying we are charging it.

All knowledge needs regular refreshing, but misconceptions need more frequent refreshing.

Archetypal questions

The classic questions for electricity involve voltage and current around series and parallel circuits. The deep structure of many of these questions is:

- In series circuits, the current is the same all the way around, but the potential differences add up to the potential difference across the cell.
- In parallel circuits, the sum of any currents entering a junction is the same as the sum of the currents leaving it.

Additional features include: At a junction, the current divides according to the resistance in the path ahead. More of the current will flow in the path with less resistance.

Before moving on to more difficult questions, make sure your students can do these.

Models

Compensating for the challenges of language and misconceptions, electricity is blessed with models. Every physics teacher has a favourite electricity model, but I think we should go for as many as possible and compare the differences.

Rope model
The rope represents the charge. Make it flow round the circuit.
Strengths:
This model clearly shows that charge is not used up in the bulbs and isn't created or stored in the battery. Resistance is simple to describe using friction at the bulbs.
Weaknesses:
It is tricky to use this model for parallel circuits. Voltage is not explained.

Water in pipes model
Instead of rope, water flows around pipes in this model.
Strengths:
This model shows what happens to current and charge at junctions, so it helps explain parallel circuits well.
Weaknesses:
I have only seen this model (and the similar 'blood and heart' and 'pizza delivery van' models) as diagrams, – requiring additional mental effort to imagine as real objects. Voltage is difficult to show in this model.

Human model
In this model, students represent discrete charges moving around a circuit.
Strengths:
Unless someone runs off, it is obvious that charge (and current) is conserved. Energy and voltage can be modelled by giving each student tokens (or sweets) as they go past the cell. The charges give their tokens up to the bulbs, modelling the transfer of energy. The cell's supply of tokens is used up, showing the battery running out of energy.
Weaknesses:
This model introduces a confusing new idea: are the people meant to be positive unit charges or electrons? If unit charges, they travel from positive to negative and each token can be a joule of energy (voltage is equal to the energy carried by each unit charge). If the charges are electrons, then they have to travel in the opposite direction and the tokens represent much smaller quantities of energy.

Lego model
In this model, the circuit is represented by Lego bricks. The height of the bricks shows the potential difference of the charge at each point in the circuit. Bulbs (and other components) are shown by slopes, demonstrating the potential difference across the component.
Strengths:
Potential difference is simple to understand as the height of the charge above ground level at a point. If charge travels through two bulbs in series, it is simple to see that the sum of the potential difference across each bulb is equal to the total potential difference across them.
Weaknesses:
This model does not show current or charge.

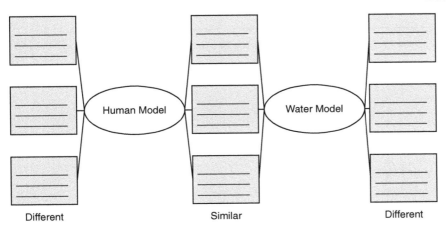

Figure 1.14 An activity for exploring similarities and differences between electricity models

Comparing the models makes strong memories. I like to use *similar/different* as a quick activity - see Figure 1.14.

Students put the properties unique to each model in the difference columns - for example, the human model has discrete charges (people), the water model has continuous charge (water) - and similarities in the central column - for example, charge flows in both models or charge can split at a junction.

Sentence models are also useful:

- *The rope model shows current by* _____,
  whereas, the water model shows current by _____
  _____
- *An advantage of the rope model over the human model is* _____

Model based reasoning

The models above allow learners to practise model based problem solving. Once the model has been learnt, you should show your students how to use it to solve problems.

Start by modelling a problem, as shown in Figure 1.15.

If the current in the ammeter on the left is 0.2A, what is the current in the ammeter in the right. Explain your answer.

The teacher thinks out loud as she models the answer to the question:

"I think about the water model. The water must be flowing around the circuit at the same speed all the way around, so the reading on the second ammeter must be the same. So the answer must be 0.2A as well. How can I write that down? Something like this:

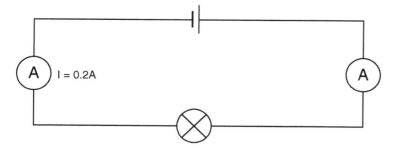

Figure 1.15 A series circuit with two ammeters

The second ammeter shows 0.2A as well, because it is a series circuit, so the current must be the same all the way around the circuit."

It is simple to vary this question by moving the ammeters and by changing the current. You could extend it by adding more bulbs and even more cells.

Goal-free activity

Circuits are also ideal for using the goal-free activity. Take Figure 1.16 for example.

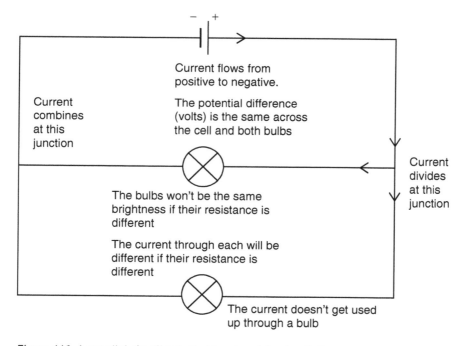

Figure 1.16 A parallel circuit labelled in a 'goal-free' activity

Practical electricity

Electric circuits are possibly the most popular experiments in the physics classroom – you just need a tray of wires, bulbs and cells. But using circuits effectively to promote learning takes a lot of thought and organisation.

Making real circuits

Circuits look easy, but they aren't. The first challenge is that the neat and tidy diagram on the board is nothing like the messy collections of wires and bulbs in a real circuit. Model to the students how to translate from the image to the real circuit and back again. This is worth practising.

Once they can interpret the diagrams and make real circuits, there are still many things that can go wrong. My list is shown in Table 1.1.

Problem	Solution
A loose connection inside the plug, so the wire doesn't work. You can't see by looking.	Have a simple circuit at the front that works (one bulb, two wires, one cell). Take out one wire from the good circuit and replace it with each wire in turn until you find the faulty one.
A bulb has blown. Sometimes you can see, sometimes you can't.	Test the bulb in the working circuit.
A bulb isn't screwed in properly.	Try screwing each bulb as securely as possible.
The wrong bulb is used. If you've got a 4.5V bulb in a single cell circuit (1.5V), it won't light up.	Look on the screw part of the bulb to check the voltage – it should be stamped in with very tiny writing.
The circuit is wired up incorrectly. Sometimes this is obvious, sometimes it isn't.	Insist that the students lay the wires out just like the circuit diagram. They can often self-diagnose.
A flat cell.	Test in the working circuit.

There are certainly more. In a real classroom, it is common to have up to half of the circuits not working and there is only one of you. Repairing one circuit might take a couple of minutes. Repairing ten circuits could take half an hour. I suggest the following:

- Test every wire and bulb before the lesson. You could train a student to do this.
- Train the students in how to test components.
- Have spare working circuits to move groups to who are waiting.

Even when circuits are working, they can still show the wrong thing! For example:

- If you connect two bulbs in series, it would be lovely if both bulbs were the same brightness. If you pick two bulbs from the tray at random, it is likely that one will be brighter than the other. Test them before the lesson if you want to be convincing.
- If you want to show that the current is the same on both sides of the bulb, you can use two ammeters, one on each side. This would be a great demonstration if both ammeters

give the same reading to 1/100 of an ampere. It is more likely that one will be a couple of digits out. This does not prove that the currents are different, it shows that ammeters are not 100% accurate. This is the same with voltmeters across parallel components. This can be a distraction to learning.

So why use circuits? The majority of problems with circuits are caused when the tray is just pulled out and expected to work. A well-prepared circuit practical gives valuable experience in the concepts of current and potential difference as well as experience solving the never-ending supply of glitches and loose connections that I am sure build character!

Van de Graaff generators

Van de Graaff generators also build character (although that isn't the main reason for using them – it is perfectly possible to use a Van de Graaff generator with no one getting a shock). They are brilliant for making voltage visible – the length of spark depends on the voltage (30,000V per centimetre of spark).

The skill of a Van de Graaff is avoiding the shock. Always earth the dome before anyone touches it (use the earth wand to do this). Don't let go and then retouch. Make sure that your leg is not close to a table or chair. Keep everyone else back.

And if you do get a shock, it isn't the end of the world – with the wrong carpet and the wrong shoes, you can get the same sized shock from a door handle. If you are not sure about a specific medical condition, get advice from CLEAPSS (2017) (http://science.cleapss.org.uk/).

Example lesson plan

Outline Plan (I have assumed a 60-minute lesson.)

This is a sample electricity lesson which includes modelling how to reason, using a physical model to explain how the current splits in series and parallel circuits. It could be used in a Key Stage 3 or 4 lesson (11- to 16-year-olds).

Time	Student activity	Teacher activity	My commentary
5 min	Starter activity: Retrieval practice (for example, KS3 Bitesize's Electric current and voltage - test (BBC) (http://www.bbc.co.uk/bitesize/quiz/q74171589)).	Sort out any problems and then monitor completion. Discreetly encourage. Quietly sanction if necessary.	*I include retrieval practice followed by spaced practice and interleaving to support development of schemata.*
	(Depending on the time available, I would only do one of the following three activities – each learner should have the activity printed so time isn't wasted copying.)		
5 min	**Option A:** Tackling misconceptions using **refutation texts** (see Student activity 1.1).	Monitor and identify best answers.	*I do this in three timed and very pacey stages. First, they write for 90 seconds, you choose three students to read their answers (make brief notes on the board), then they have 60 seconds to rewrite. Repeat process for second text.*

5min	**Option B:** *Solo-pair-share* **goal-free** task (see Student activity 1.2).	Monitor during the solo and pair stages, identifying the best ideas to bring out in the share stage.	*Again, I use a timer to keep the pace high – probably 60 seconds each for the solo and pair stages. Keep tight control of the share stage. Monitor whether individuals are updating their sheets.*
5 min	**Option C:** *Solo-pair-share* **elaboration task** (see parallel and series circuit similar/different sheet in Student activity 1.3).	Monitor during the solo and pair stages, identifying the best ideas to bring out in share stage.	*Again, I use a timer to keep the pace high – probably 60 seconds each for the solo and pair stages. Keep tight control of the share stage. Monitor whether individuals are updating their sheets.*
35 min	Practical: Series and parallel circuits model (see Student activity 1.4).	See below.	See below.
10 min	Review/exit-ticket.	Prepare series and parallel circuits questions with current values to solve.	You could model the first one and then ask the students to complete.

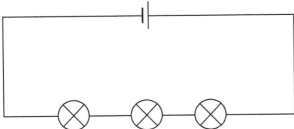

Figure 1.17 A series circuit with three bulbs

Student activity 1.1: Option A - Refutation texts

1 Some people think that current gets 'used up' by a bulb, but _____.
2 Some people think that all the bulbs in a series circuit will be the same brightness, but _____.
3 Some people believe that the current in both bulbs in a parallel circuit will be the same, but _____.

Student activity 1.2: Option B - Goal-free

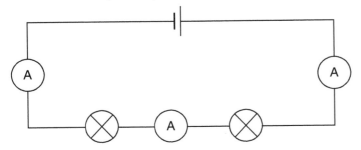

Figure 1.18 A series circuit with three ammeters

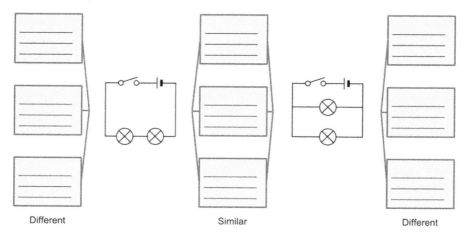

Different   Similar   Different

Figure 1.19 An activity for exploring similarities and differences between parallel and series circuits

Student activity 1.3: Option C - Elaboration

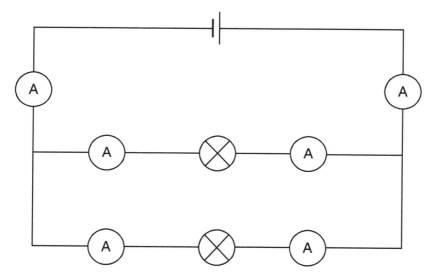

Figure 1.20 A parallel circuit with six ammeters

Student activity 1.4: Series and parallel models practical

Set up two circuits - one series with two bulbs, the other parallel with two bulbs. In both circuits, put ammeters either side of the bulb as in Figures 1.20 and 1.21. Before you switch the circuits on, ask students to predict what they will observe about the currents (in pairs). Switch on the circuits and record the currents.

Now use a human model to explain the results: the students are the charges moving round the circuit, you can be the cell and ask two students to be a bulb and four students to be ammeters.

Use masking tape on the floor of a large empty area, mark out the series and parallel circuits.

The students who are charges stand on the tape 'wires' of the series circuit and move in the direction you (the cell) specify. As charges pass the bulbs, the bulbs light up (wave).

The ammeters count the number of charges passing them in one minute (the current). Record the results.

Repeat for the taped parallel circuit.

Your task is to demonstrate how the model corresponds to the real circuits. Think out loud as you write this:

> In the series circuit model, all the charges had to go all of the way around, which explains why the current was the same all of the way around in the real circuit.

> In the parallel circuit model, all the charges went through the cell, so that had a large current, which is what we saw in the real circuit. When the charges got to the first junction, they had to split. Some went one way and some went the other, but the total number of charges was the same. We could see this in the real circuit because the two currents after the junction combine to equal the current before the junction. At the second junction, the charges recombined. This explains why the current returning to the cell was the same as the current leaving it.

Once you have modelled the text, think about what you want them to practise. You might decide to give sentence starters for them to complete.

You might remove your text and ask them to write their own version.

You might ask them to practice reading the texts aloud to each other in pairs so that they become familiar with the sentence structures used (quite tricky ones).

Conclusion

Electricity is the perfect topic to practise teaching physics. It is full of abstract knowledge, which we understand through models; the practical work is relatively straightforward, but needs practice to master; the vocabulary is both vague (for example 'electricity') and precise (for example 'potential difference') and there are plenty of misconceptions to overcome.

Use a narrative wherever you can to jump-start the schema. Model effective answers and quiz as much as you can.

Further reading - Electricity

The following Institute of Physics (IoP) sites are useful:

- www.practicalphysics.org/electric-circuits-and-fields.html has comprehensive instructions for teaching a wide variety of practical activities.
- http://supportingphysicsteaching.net/ElHome.html.
- www.talkphysics.org hosts a community of physics teachers which I find very helpful. Experienced colleagues answer questions and share advice, usually very quickly.

In addition, I recommend:

- MOSART (Misconceptions-Oriented Standards-Based Assessment Resources for Teachers), www.cfa.harvard.edu/smgphp/mosart/.
- *Physics for You*, Johnson, Oxford University Press, 2016.

2 Forces at a distance

Action at a distance: Peregrinus, Gilbert, Newton, Faraday, Maxwell, Einstein and Fermi.

Students find forces at a distance mysterious and confusing. It helps them to see that the best physicists have struggled with the same problems. Telling the stories behind the development of *fields*, *lines of force*, *curved space* and *force carrying particles* exploits the brain's ability to make sense of stories. The narrative provides the beginning of a schema for phenomena that are mysterious.

> Spooky action at a distance.
>
> Albert Einstein

It is unsettling to see objects move without being touched. Children (all of us?) love to make a magnet move on a table by moving another magnet underneath. There is magic in it.

Philosophers have always struggled with invisible forces acting without touching. Gravity does it, as does magnetism and electricity. In the 16th and 17th centuries, philosophers described these forces as the *soul* of the object: magnetism is the *soul* of the Earth, gravity the *soul* of the universe. Forces which act without any physical contact were thought *occult*: mysterious and spiritual.

Some great physicists at the time tried to distance themselves from this way of thinking. In the 17th century, Gottfried Leibniz redefined *occult* as simply *unknown*.

> The ancients and the moderns, who own that gravity is an occult quality, are in the right, if they mean by it that there is a certain mechanism unknown to them whereby all bodies tend towards the center of the earth.
>
> (Leibniz quoted by Newton 2004: 151)

Isaac Newton simply ignored the mystery in his Law of Universal Gravitation. When philosophers objected, saying that it did nothing to explain the mystery, only calculate, Newton (1729: 393) refused to be drawn. He wrote: "I frame no hypothesis."

Frame. Like a criminal.

But the idea of the soul of the universe causing the motion of objects has been very difficult for physicists to resist. It is a medieval way of thinking. A human way. Our understanding of forces acting at a distance has always been a struggle between the rational and the occult. It probably always will be.

Petrus Peregrinus, Crusader - 1269

The first modern account of magnetism was written during the crusades. On the 8th of August 1269, Petrus Peregrinus, a soldier and engineer, wrote a letter to a friend. The letter contained an astonishing description of the properties of magnets:

> "Know then that this is the law: the north pole of one lodestone attracts the south pole of another, wile the south pole attracts the north. Should you proceed otherwise and bring the north pole of one near the north pole of another, the one you hold in your hand will seem to put the floating one to flight. If the south pole of one is brought to the south pole of another, the same will happen.
>
> (Peregrinus 1269: 11)

In other words: like poles repel, unlike poles attract.

A lodestone is a naturally magnetic rock - when you hang it on a thread, it will always turn so that the same side points north and the other points south. This magical property was known to the ancient Chinese, and the Vikings used lodestones to navigate at sea. Peregrinus had two.

Laying siege to a city, like all soldiering, has terrifying moments of activity punctuating long periods of tedium. Peregrinus kept himself busy experimenting (he wanted to build a perpetual motion machine powered by lodestones). When he brought his two lodestones close to each other, they no longer pointed north. Slowly they turned to face each other, one north-facing side always pointing to the other's south-facing side.

Peregrinus' letter was copied and distributed throughout Europe. Thirty-nine copies of the letter still exist - handwritten on vellum.

Three centuries after it was written, a copy made its way to Colchester, to an English gentleman's desk.

William Gilbert of Colchester, Physician to Queen Elizabeth I - 1600

When he wasn't attending to the queen, William Gilbert spent much of his effort discrediting ancient science. He was a very angry man:

> These are the utterly false and disgraceful tales of the writers.
>
> (Gilbert 1600: 48)

He sat at his table with his Greek texts (and a copy of Peregrinus' letter) and set about putting the record straight. His book, *De Magnete*, does this brilliantly.

Gilbert studied both magnetism and electrostatic force. He invented two marvellous devices to do this.

His first great invention was called the versorium - a metal pin that could freely rotate, like an un-magnetised compass needle. When a rubbed object, such as amber, is brought near the needle, it rotates and points towards the object. Whereas the Greeks believed that only rubbed amber could attract, Gilbert used his needle to prove many other materials could also attract, including diamond, sapphire, glass, sulphur, sealing wax and resin. He named these materials *electrics* because they behave like amber (elektron).

It is enjoyable to note that despite Gilbert's great outrage and hubris, he made his own scientific errors. Because his needle could only attract, he believed that electricity could not repel. It is simple to prove otherwise: if he had used Peregrinus' trick of hanging two charged stones next to each other, he would easily have seen them repel.

Gilbert's second great idea was that the Earth was a magnet. To prove it, he made his own model Earth: a lodestone sphere with the south-pointing pole at the top and the north-facing pole at the bottom. He called it his Terella.

With the Terella, Gilbert was able to demonstrate how a compass will always point north. He could also reproduce an effect familiar to navigators – a compass stops working in the far northern latitudes. Finally, he was able to show why a compass needle suspended on a thread will dip downwards – the amount of dip depending on the latitude. This was enough to convince Gilbert that the Earth is a giant magnet.

Gilbert also believed that magnetism was the soul of the Earth. No matter how rational Gilbert was, he was still a man of his time.

Newton's Law of Universal Gravitation – 1687

At the start of the 17th century, our understanding of gravity was far behind electricity and magnetism. By the end, it was far ahead.

In 1630, thirty years after Gilbert, Johannes Kepler, the astronomer, was struggling with the idea of the soul of the universe:

> For I once believed that the cause moving the planets was entirely a soul. But when I con-sidered that this moving cause weakened with distance, and that the sun's light too is atten-uated with distance from the sun, I came to the conclusion that this is some kind of force.
>
> (Kepler 1630 quoted by Hetherington 1993: 351)

There was no theory of gravity for Kepler; instead he used Gilbert's magnetism to explain the motion of the planets. By the end of the 17th century, Newton had published his Law of Universal Gravitation – a law that successfully described the motion of the universe for the next 300 years.

Newton's Law of Universal Gravitation seemed to be the product of one brilliant mind. It is a complete law governing the motion of all celestial bodies in the universe. We use it today to calculate the paths of spacecraft to the planets, to calculate the orbits of plan-ets around distant stars and to predict the motion of galaxies. But there were two minds: Newton's and Hooke's.

Isaac Newton and Robert Hooke had a difficult relationship. Hooke believed that Newton failed to acknowledge his role in the development of our understanding of gravity. Newton resented this accusation, and Newton bore grudges. When Hooke died, Newton worked hard to remove his legacy. Hooke's portrait and all of the equipment donated by Hooke to the

Royal Society disappeared, with many historians accusing Newton of destroying them. There is no mention of Hooke in any of Newton's publications.

But Hooke and Newton had discussed gravity together. In 1679, Hooke wrote Newton a letter. Hooke believed that planetary orbits were straight lines combined with continuous deflections from the line. It was a brilliant place to start a discussion. Hooke was asking for help.

A short series of letters began between the two men, each bringing them closer to understanding gravity. Hooke provided the idea that all celestial bodies have an "attraction towards their common centre" (The Philosophical Transactions 1747: 286). He wrote that deflections from a straight line would cause orbits to be circular or elliptical.

Newton's reply included a thought experiment. He imagined a tall church with no draughts and someone holding a pistol shot high inside the church. What would happen as the ball dropped? Newton said that most people ('the vulgar') would suppose that the ball would fall slightly to the west, because while the ball was falling, the Earth would rotate a very small amount. In fact, said Newton, the opposite was true: the ball would fall towards the east. This is because every object on the rotating Earth is carried at a velocity. An object with a greater height has a higher velocity. The ball at the top of the church has a higher velocity than the ground at the bottom. As the ball falls, its greater sideways velocity causes it to overtake the ground, moving slightly eastwards.

This is correct. But then Newton did something remarkable: he imagined the impossible. He imagined the ball continuing to fall through the ground towards the Earth's centre and asked how the ball would continue to fall. Newton said that the ball would fall in a spiral.

This was wrong and Hooke knew it. He took Newton's letter and read it aloud at the Royal Society, pointing out Newton's errors. Hooke said the ball would take a circular or elliptical path, orbiting the centre of the Earth like a planet.

Despite this unkind and ungenerous act, the letters continued, with each man developing his ideas of the falling ball. Finally, Hooke wrote the line that he later claimed was at the heart of Newton's Law of Universal Gravitation (see Figure 2.1):

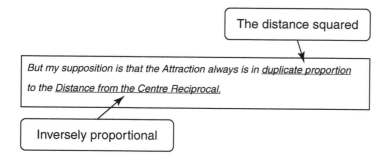

Figure 2.1 Close reading of Hooke's quote

But my supposition is that the Attraction always is in duplicate proportion to the Distance from the Centre Reciprocal.

(Hooke 1679 quoted by Westfall 1994: 152)

In other words, gravity obeys an inverse square relationship – the idea at the heart of Newton's Law of Universal Gravitation. Hooke finished his letter with a request for Newton to help him solve the maths and find the shapes of orbits: "I doubt not that but by your excellent method you will easily find out what that Curve must be" (Hooke quoted by Whiteside 2008: 13). Newton did not reply.

Four years later, in a coffee shop in London, Robert Hooke, Christopher Wren and Edmund Halley were debating gravity. Hooke asserted his inverse square relationship. Wren proposed a challenge: he would give 40 shillings to whichever of Halley or Hooke was able to calculate gravity mathematically. Neither man was successful.

In August that year, Halley was visiting Newton in Cambridge. He asked Newton what shaped path a planet would take if gravity obeys an inverse square rule. Newton replied that it would be an ellipse and claimed that he had the calculations in his rooms, but was unable to produce them. Nevertheless, by November Newton had published a manuscript (now lost) called *De motu corporum in gyrum (On the motion of bodies in an orbit)*. In it he presented the inverse square law of gravitation and showed how this leads to elliptical orbits. There was no mention of Hooke's contribution.

Newton took the idea of gravitation further than this. He understood that when the moon's path is bent from a straight line into an ellipse, it must be accelerating. He imagined a little moon orbiting the Earth at a height just above the mountains. Using Kepler's law, the time of the orbit of the little moon must be about 90 minutes. The acceleration needed to cause this is $9.8 m/s^2$, exactly the same as the acceleration of a ball dropped. He concluded that the force causing a ball to fall and the force holding the moon in its orbit must be the same.

But Newton could not provide a mechanism for gravity. He wrote: "Gravity must be caused by an Agent acting constantly according to certain Laws; but whether this Agent be material or immaterial, I have left to the Consideration of my Readers" (Letter to Richard Bentley in Newton 1756: 26).

One such reader was Michael Faraday.

Faraday's lines of force – 1852

By the 18th century, philosophers had abandoned the idea of *soul* to explain forces acting at a distance. But they still craved explanation. Michael Faraday couldn't let it go:

> In the numerous cases of force acting at a distance, the philosopher has gradually learned that it is by no means sufficient to rest satisfied with the mere fact, and has therefore directed his attention to the manner in which the force is transmitted across the intervening space.
>
> (Faraday 2012: 408)

Michael Faraday's great contribution to the study of forces at a distance was to imagine the space between the objects – the lines of force, or field, as shown in Figure 2.2.

At first, Faraday proposed these lines as a model to help us visualise the forces. Like contour lines on a map, they were useful, but not there in reality. Gradually his language changed. Faraday came to believe in the lines.

Secretly, most of us agree: we talk as though the lines are real. When we talk about a wire cutting the lines of magnetic flux in a dynamo, we imagine the wire passing through

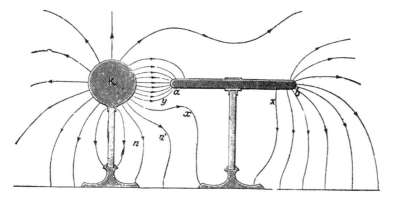

Figure 2.2 Faraday's lines of force

individual lines. When the lines are closer together, and the wire cuts through them more quickly, the induced voltage is greater.

The lines of force aren't occult, but they are mysterious.

Maxwell's equations: The second great unification in physics - 1865

Faraday's lines of force, or field, were a fundamental development from Newton's idea of the universe. No longer was the universe an empty stage for particles to act upon. Instead it was filled with fields and lines of force. Newton unified mechanics and astronomy with his Law of Universal Gravitation: one equation to explain the whole of gravity. Faraday did not write the equations to unify electricity and magnetism; this was done 20 years later by James Clerk Maxwell.

In 1865, James Clerk Maxwell published his great work: *A Dynamical Theory of the Electromagnetic Field*. In it, he does for electricity and magnetism what Newton did for mechanics and astronomy: he combines two separate branches of science into one.

To do this, Maxwell uses four equations, as shown in Figure 2.3.

$$\nabla \cdot E = \frac{\rho}{\varepsilon_0}$$

$$\nabla \cdot B = 0$$

$$\nabla \times E = -\frac{\partial B}{\partial t}$$

$$\nabla \times B = \mu_0 J + \mu_0 \varepsilon_0 \frac{\partial E}{\partial t}$$

Figure 2.3 Maxwell's equations

These four equations require maths not typically taught until university. But even without the maths, we can understand what they mean.

The first equation shows that electric fields, or lines of force, emerge from positive charges and vanish into negative charges.

The second equation shows that magnetic fields always form loops - they do not have a beginning or an end. We see this in diagrams about magnetic fields through coils: the lines of force make complete loops. In solid magnets, it looks as though the lines of force start at the north and end at the south poles, but Maxwell's equation shows that these fields travel through the solid magnet to make complete loops.

The third equation shows that a changing magnetic field will cause an electric field. When you move a magnet near a wire, or move a wire near a magnet, a voltage will be induced. This is how dynamos and generators work.

The final equation shows us several phenomena. If the current is constant (for example in a simple circuit), you get a magnetic field (for example an electromagnet). But if there are no charges, a changing magnetic field will produce a changing electric field, which produces a changing magnetic field, which produces a changing electric field, and so on. In other words, an electromagnetic wave produces a light wave! Maxwell discovered that light is an electromagnetic wave. More than that, this equation lets you calculate the speed that light travels in different materials.

Maxwell's contribution to physics is huge and he deserves to be remembered alongside the greats.

Einstein's curved space - 1915

Newton's Law of Universal Gravitation describes the forces acting through space. Everything acts in straight lines and behaves mathematically. But when he ignored the question of how gravity worked, he upset philosophers. Einstein put that right when he published the theory of General Relativity, described as the most beautiful of theories. Einstein found a very different explanation to Faraday's lines of force.

Einstein's great revelation was that the gravitational field does not pass through space - he realised that the gravitational field and space are the same thing. He didn't need forces acting at a distance to explain how objects are affected by gravity - they accelerate because space is curved around massive bodies: no field lines, just curved space.

Fermi's nuclear forces - 1933

And now we have another way of understanding forces acting across empty space (although the distances for these new forces are tiny: the size of a nucleus). Just seventeen years after Einstein published his paper on General Relativity, and unexpected by the physics establishment, Enrico Fermi proposed two new forces.

In 1933, Fermi tried to publish a letter to explain how subatomic particles behave. He described the two new forces able to explain nuclear phenomena: the weak nuclear force and the strong nuclear force. The journal *Nature* refused to publish his paper for five years because it contained "abstract speculations too remote from physical reality to be of interest to the readers" (Bernardini and Bonolis 2013: 346).

Fermi's forces did not use *lines of force*. He used particles to carry the forces between subatomic particles. Today we call these particles *gauge bosons*. They carry enormous forces over tiny distances, holding the particles of a nucleus together.

The history of forces teaches us that the deepest and most useful physics is full of *abstract speculations too remote from reality*. When we teach forces at a distance and pretend that we understand how forces act without touching, we are not being honest. It is a deeply mysterious subject and, as Einstein said, "spooky."

Teaching forces at a distance

We can't hope to understand the abstract universe of fields, curved space and force-carrying particles. In fact, the word *understand* does not feel right – we learn to use the ideas until we are comfortable with them. So, as with so much of physics, practice comes before understanding. For most of us, there is no "aha!" moment – just a growing sense of competence and confidence. The rest of this chapter suggests ways to accelerate this process for your students. Start by fixing the foundations: your students' misconceptions.

Misconceptions

Magnets

The first great misconception that many students have to learn to overcome is one that they learn by mistake – most children *learn* that all metals are magnetic. I have never seen teachers confuse this: they are very clear about the types of metal, demonstrating that copper and aluminium are not attracted but that iron and nickel are. Still the misconception arises. A key part of learning is forgetting; we simplify and categorise. *Iron is magnetic* turns into *some metals are magnetic*, which turns into *metals are magnetic*.

Later this confusion can return when students learn about electromagnetism –that they must use a *soft iron core*. Soft iron loses its magnetism very quickly, which is important in electromagnets using alternating current. Many students find it difficult to remember about the soft iron core – they can't fit it into their schema because they suspect their schema is wrong.

It is also common for students to believe that the wires in an electromagnet need to be made from iron or nickel. In this case, however, the metal is simply acting as a pathway for the electric current to flow; a beam of electrons (a *cathode ray*) has a magnetic field without any conductor at all.

The magnetic fields around currents are abstract. Any misconceptions are likely to have been learnt in a science lesson. The key things to know about magnetic fields are:

- The arrows point from north to south.
- The magnetic field is strongest where the lines of force are closest together (usually near the poles).
- Magnetic field lines cannot cross because a force acting on a particle can only have one direction; a point where a field line crosses would have two directions. This is true for all fields.
- You cannot separate a north pole from a south pole – if you cut a magnet in half you will have two smaller magnets.

Electrostatics

When learners try to link their electricity and electrostatic schemata they often become confused. Electrostatics work on insulating materials (for example paper, hair and dried herbs) as well as conducting materials (for example small pieces of metal foil and water). But when they learn about current, only conductors work.

Making the differences explicit often helps, as shown in Figure 2.4.

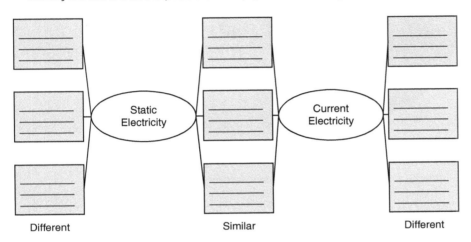

Figure 2.4 An activity for exploring similarities and differences between static and current electricity

Sentence models reduce cognitive load, for example:

- In static electricity, the voltage is typically _____, whereas in current electricity _____.
- In static electricity, the current is typically _____, whereas in electric circuits _____.
- In circuits, the charges _____, whereas in static electricity _____.

Gravity

GRAVITY IS DOWN

Hopefully your students will have the idea that the Earth is a sphere and that gravity is a force which acts towards the centre of mass. But this instinctive misconception that gravity is down can lie alongside the correct version for years. To assess how well the misconception has been buried (it never truly disappears and may spring back to life when you are unwary or stressed), you can adapt the archetypal question – 'which direction will the apple fall when dropped?' – by replacing the Earth with another body, for example the Moon, Mars, Jupiter, the Sun or an irregular shaped object such as an asteroid. You will have to patiently fix each one (see Figure 2.5).

Figure 2.5 Activity for students to draw in the direction of the force

THERE IS NO GRAVITY IN SPACE

This misconception arises either because we see astronauts in orbit floating weightlessly or because of confusion about no atmosphere meaning no gravity. But if there was no gravity, the astronauts (and their space-station) would shoot off into space. Instead, the astronaut and the space-station are orbiting together, which makes the astronaut look and feel weightless.

You can reason that the Sun's gravity must extend at least as far as Pluto, because Pluto orbits the Sun.

The extension to this misconception is that there is no gravity on the Moon (or other planets). Obviously there is, otherwise all of the rocks would fall off.

Example refutation sentences you could use include:

- Many people believe that there is no gravity in space, but there must be, otherwise the planets would shoot off into space.
- Many people believe that there is no gravity in space, but it just looks like that when objects and astronauts float around inside the space-station (in reality, the astronauts, objects and space-station are all orbiting together).
- Many people believe that there is no gravity on the Moon, but there must be, otherwise all the rocks would fall off.

Archetypal questions

Classic questions for this topic include:

Magnets

- Draw the magnetic field around a current-carrying wire, coil or solenoid.
- Recall how to increase the strength of an electromagnet.

- Explain how a transformer works.
- Describe how to induce a current in a coil by moving a bar magnet through it.
- Describe the motor effect.
- Explain how a motor/speaker works.

Electrostatics

- How can vehicles reduce the risk of sparks when fuelling?
- How do clouds become charged?
- How does a lightning conductor work?
- How does a laser printer/photocopier work?
- How is static electricity used to spray paint on cars?

Gravity

- Use Newton's Law of Universal Gravitation to explain why the orbit of Pluto takes longer than the orbit of Mercury.
- Explain why a satellite does not fall to the Earth, or fly off into space.

Using strategies from cognitive psychology in lessons

Quizzes - Retrieval practice

If you've taught something, it's wasteful to let that knowledge fade. Make sure your students maintain what they've learnt by frequent practice. Quizzes are great for this.

BBC Bitesize has appropriate quizzes and there are many more available. I use quizzes as part of a starter activity in most of my lessons as they are great for retrieval practice (see Sumeracki and Weinstein (n.d.) on retrieval (www.learningscientists.org/retrieval-practice)). Multiple choice questions are very quick to mark in class. (I include questions from other physics topics, to make use of the interleaving effect – see Sumeracki and Weinstein (n.d.) on interleaving (www.learningscientists.org/interleaving)).

There is nothing wrong with repeating the same quiz many times, as long as the questions are about knowledge you really want your students to remember.

Gap-fill

Gap-fill activities can reduce cognitive load, which leads to better learning, but take care: sometimes the cognitive load is increased because the meaning of the sentence is not obvious without the missing word – effort is wasted working out what the question means. For example:

1   When two magnets are brought up to each other, north pole to north pole, they will
_____.
2   The magnetic field is _____ near the ends of the magnet.
3   Aluminium and copper are not _____ by a magnet.

Goal-free strategy

The goal-free strategy reduces cognitive load by asking the learner to write as much as she can about the diagram. It works well as part of a *think-pair-share* activity. Figure 2.6 is a completed example.

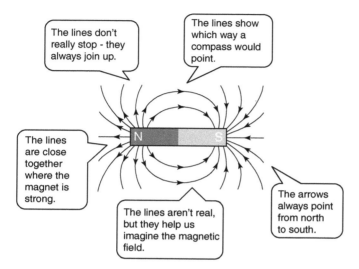

Figure 2.6 An example of a completed 'goal free' activity where learners add as much information to an unlabeled diagram as they can

Elaboration

Another effective learning strategy from cognitive psychology is elaboration (see Sumeracki and Weinstein (n.d.) on elaboration (www.learningscientists.org/elaboration)). Elaboration works by encouraging learners to explicitly make comparisons between schemata. The example in Figure 2.7 encourages learners to compare and contrast a gravitational field with a magnetic field. The learner's task is to write down as many similarities and differences as she can recall.

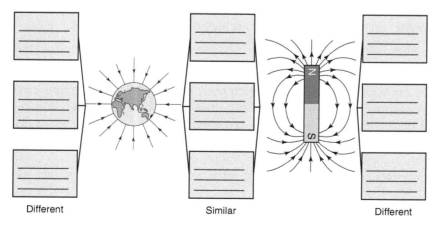

Figure 2.7 An activity for exploring similarities and differences between a gravitational field and a magnetic field

I find it helpful to ask students to use the collaborative strategy *think-pair-share* to gather the best ideas from the class.

Using demonstrations and practical work for writing

One of the key uses for practical demonstrations in physics lessons is to provide a prompt for thinking. Writing is a key tool for slowing thinking down. I have described several writing activities below which reduce cognitive load, supporting problem solving and learning.

The suspended magnet

Recreate Peregrinus' experiment with lodestones using two bar magnets. Suspend one bar magnet horizontally on a thread from a retort stand. Bring the other close to it. Make and record the following observations:

- *When north is brought close to north, _____.*
- *When north is brought close to south, _____.*
- *When south is brought close to south, _____.*

Your students are practising:

- The useful 'when xxx, yyy' sentence type.
- The phrase 'brought close to'.
- The idea behind 'like poles repel, unlike poles attract', the most succinct and elegant sentence in physics.

The versorium

Your students may be asked to write an explanation of how a gold-leaf electroscope works. This has a very high cognitive load. The versorium needle is an earlier and simpler version of the electroscope which makes it an ideal starting point.

The versorium needle is quick and simple to make: cut a long, thin diamond shape out of aluminium foil (about 10cm by 4cm). Fold it along the long line of symmetry. Open the folded shape slightly. Balance your foil needle on a cocktail stick (use plasticine to hold the cocktail stick vertical) so that the foil needle can rotate like a compass.

Charge an insulating material (such as a pen, ruler, glass rod, comb, etc.) by rubbing it. Bring it close to the needle. It will attract, regardless of whether the charge on the object is positive or negative.

The versorium can be used as a precursor experiment to teach electroscopes. When you bring a positively charged object up to the needle, electrons in the needle are attracted closer to the object. This negative end of the needle is then attracted to the charged object.

When a negatively charged object is brought close to the needle, electrons are repelled to the far end of the needle, leaving a positive charge on the end closest to the negative object, which attracts the needle. In other words, the needle always attracts.

This concept has a high cognitive load. You can reduce it by ensuring the following are already well learnt:

- Positive attracts negative.
- Negative repels negative.
- Positive repels positive.
- Electrons are negative and are free to move.

You can break down the writing into several sentences:

1   When a *negatively* charged object is brought close to the needle, electrons in the needle _____.
2   The end of the needle closest to the *negatively* charged object now has a _____ charge.
3   The needle and the object now _____.

You can ask your students to translate the three sentences above by replacing *negatively* with *positively*.

Model the text for your students to: *explain why the needle is attracted to a positive object.*

Your students should now be ready to write the text to *explain why the needle is attracted to a positive object.*

Gold-leaf electroscope

Historically and cognitively, the gold-leaf electroscope is a development of the versorium and works in the same way. This time, however, the gold-leaf is always repelled because the induced charge on the leaf and the induced charge on the metal bar it is attached to are always the same. You can use a similar strategy to The versorium writing task.

Why not finish off with a *similar/different* activity comparing the versorium needle with the gold-leaf electroscope?

Suspended, charged nylon and glass rods on threads

Recreate Abbe Nollet's experiment demonstrating two types of electricity (see Chapter 1: Electricity).

1   Charge one glass rod by rubbing with a silk cloth (vitreous electricity in Nollet's terms, positively charged in ours). Suspend it horizontally on a thread attached to a retort stand.
2   Repeat with a nylon rod rubbed with fur (resinous electricity in Nollet's terms, negatively charged in ours).
3   Charge a second glass rod and bring it close to each of the two suspended rods.
4   Charge a second nylon rod and bring it close to each of the two suspended rods.

To reduce the cognitive load, sentences starters can be used as a completion exercise. For example:

- *When the positively charged glass rod is brought near to a similarly charged glass rod, _____.*
- *When the positively charged glass rod is brought near to the negatively charged nylon rod, _____.*
- *When the negatively charged glass rod is brought near to a similarly charged nylon rod, _____.*
- *When the negatively charged glass rod is brought near to a positively charged glass rod, _____.*

(*Note:* It is worth emphasising the phrase 'similarly charged' – it is a useful phrase that you want your students to recognise and use. The sentence format is also useful: 'when xxx, yyy.')

Example lesson plan

Outline plan (I have assumed a 60-minute lesson.)

This is a sample lesson which includes a practical demonstration. The aim is to practice explaining how a gold-leaf electroscope works. It is important that students understand how an electroscope works before they carry out any other electroscope experiments (for example see Institute of Physics (2007) on a class practical for charging electrostatic induction (www.practicalphysics.org/charging-electrostatic-induction.html)).

I have assumed that the students have already been taught to explain how the versorium needle works (see the versorium writing task on [page]).

Time	Student activity	Teacher activity	My commentary
5 min	Starter activity: Retrieval practice (see Student activity 2.1).	Sort out any problems and then monitor completion. Discreetly encourage. Quietly sanction if necessary.	I've included retrieval practice followed by spaced practice and interleaving to support development of schemata.
	(Depending on the time available, I would only do one of the following three activities – each learner should have the activity printed so time isn't wasted copying.)		
5 min	**Option A:** Tackling misconceptions using **refutation texts** (see Student activity 2.2).	Teacher monitors and identifies best answers.	I'd do this in three timed and very pacey stages. First they write for 90 seconds, you choose three students to read their answers (make brief notes on the board), then they have 60 seconds to rewrite. Repeat process for second text.
5min	**Option B:** *Solo-pair-share **goal-free** task* (see Student activity 2.3).	Teacher to monitor during the solo and pair stages, identifying the best ideas to bring out in the share stage.	Again, I'd use a timer to keep the pace high – probably 60 seconds each for the solo and pair stages. I'd keep tight control of the share stage. Monitor whether individuals are updating their sheets.

Time	Student activity	Teacher activity	My commentary
5 min	**Option C:** *Solo-pair-share similar/different task* (see Student activity 2.4).	Teacher to monitor during the solo and pair stages, identifying the best ideas to bring out in share stage.	Again, I'd use a timer to keep the pace high – probably 90 seconds each for the solo and pair stages. I'd keep tight control of the share stage. Monitor whether individuals are updating their sheets.
25 min	Model the gold-leaf electroscope (see Student activity 2.5).	Before you begin, read the Institute of Physics' very useful text on electrostatics(www.practicalphysics.org/electrostatics.html) Demonstrate how a gold-leaf electroscope can show charge by bringing a charged object to the top of the electroscope without touching. Use the model described in Student activity 2.5. Model a text explaining what happens when a positive rod is brought towards the plate. Provide sentence starters for your students to complete for the negative rod.	
10 min	Reading: Learn how to use the electroscope from the Institute of Physics (http://practicalphysics.org/using-electroscope.html).	Prepare access to the text. Discreetly encourage. Quietly sanction if necessary.	The 'Charging by induction' section of the article is most relevant. I would ask students to read this text aloud. If you have enough gold-leaf electroscopes, they can follow the instructions given in the text.
10 min	Review/exit-ticket (see Student activity 2.6).	Prepare *similar/different* activity for the gold-leaf electroscope and the versorium.	

Student activity 2.1: Starter quiz

1   What does the phrase 'like charges repel' mean?

   a   A positive charge feels a force away from another positive charge
   b   A negative charge feels a force away from another negative charge
   c   Both of the above.

2   Which charges are free to move in a conductor?

   a   Electrons (negative charge)
   b   Positive metal ions
   c   Electrons and positive metal ions.

3   When two different insulators are in contact (e.g. rubbed together):

    a   Charge moves from one insulator to the other

    b   Positive charge is created on one insulator and negative charge is created on the other

    c   Both of the above.

4   When a charge is induced:

    a   Charge is created out of nothing

    b   Charge flows from one place to another without any contact

    c   An object gets charged by rubbing.

Student activity 2.2: Option A – Refutation texts

Some people get confused between a compass needle and the versorium needle – the compass needle _____ but the versorium needle _____ .
Many people think that positive charge moves in a metal, but _____ .
Some people think that charge is given to the versorium needle when a charged object is brought close, but _____ .

Student activity 2.3: Option B – Goal-free

Give your students an unlabelled diagram of the versorium needle to label with as much information as they can. Use *solo-pair-share*: give students 60 seconds to label as much as they can individually and then 60 seconds to work in pairs or threes to add as much as they can. While they are doing this, you can assess students discreetly and at the same time iden-tify two or three valuable contributions to share with the whole class.

Student activity 2.4: Option C – Similar/different

This activity is good at highlighting the differences between magnetism and electrostatic behaviour, using Figure 2.8 as a prompt.

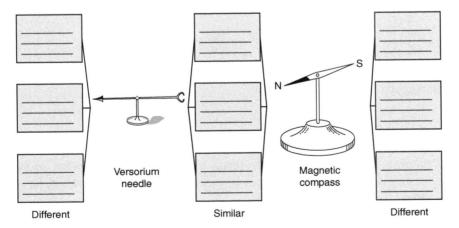

Figure 2.8 An activity for exploring similarities and differences. between the versorium needle (detecting electrostatic fields) and a magnetic compass needle

Key differences:

- The versorium needle is not magnetised, it has induced charge when another charged object is brought near.
- The charge at the point of the versorium needle is the same as the charge on the object brought near to it.
- The compass needle is magnetised and reacts to other magnetic fields.
- The compass needle can attract or repel.
- The polarity of the compass needle does not change – north stays north.

Use *solo-pair-share*: give students 90 seconds to label as much as they can individually and then 90 seconds to work in pairs or threes to add as much as they can. While they are doing this, you can assess students discreetly and at the same time identify two or three valuable contributions to share with the whole class.

Student activity 2.5: Model the gold-leaf electroscope

Draw a diagram of a gold-leaf electroscope onto a whiteboard or flipchart, using Figure 2.9 as a guide. Draw positive metal ions ('+') covering the gold-leaf, the metal stem and the plate at the top.

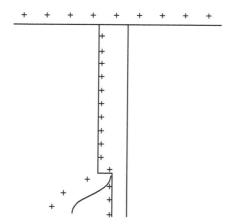

Figure 2.9 Diagram of a gold-leaf electroscope

For every metal ion, add a small sticky note with a negative ('–').

Use the model to demonstrate what happens when you bring a *negatively* charged rod down towards the plate at the top of the electroscope: the negative charges on the sticky notes are repelled down into the leaf and the stem – the leaf and the stem now obviously have extra negative charge, so they repel.

Use the model to demonstrate what happens when you bring a *positively* charged rod down towards the plate at the top of the electroscope: the negative charges on the sticky notes are attracted up into the plate, leaving the leaf and the stem positively charged - the leaf and the stem repel.

Demonstrate this effect again using the real electroscope.

Model the text:

> When you bring a positively charged rod towards the plate, the negative electrons are repelled down into the stem and the leaf. The leaf and the stem are now both negatively charged, so they repel.

Now get the students to write the version for a negatively charged rod. You can provide these sentence starters: to reduce the cognitive load.

> When you bring a negatively charged rod towards the plate, _____.
> The leaf and the stem are now both _____, so _____.

Student activity 2.6: Review/exit-ticket - Similar/different

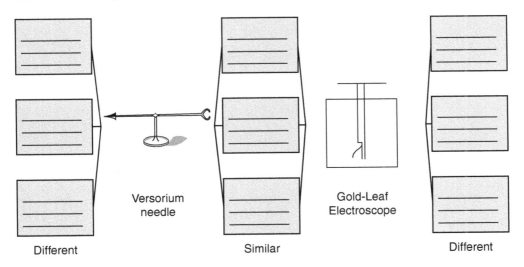

Figure 2.10 An activity for exploring similarities and differences between the versorium needle (detecting electrostatic fields) and a gold-leaf electroscope

Conclusion

Forces at a distance are spooky, mysterious and abstract. They have changed how we understand space: a web of field lines carrying light alongside forces, a flexible grid of time and space distorted by huge masses and a subatomic world fizzing with force-carrying particles. We make sense of this world by answering questions until it starts to feel normal. Start by telling the stories - we are used to magical worlds. Then show your students how to answer the questions, gradually reducing the support until they become independent. Very few students can do this alone - your task is to be their guide and mentor.

Further reading - Forces at a distance

The following Institute of Physics (IoP) sites are useful:

- www.practicalphyiscs.org has comprehensive instructions for teaching a wide variety of practical activities.
- http://supportingphysicsteaching.net/RaHome.html.
- www.talkphysics.org hosts a community of physics teachers which I find very helpful. Experienced colleagues answer questions and share advice, usually very quickly.

In addition, I recommend:

- *Children's Ideas in Science*, Driver, Guesne and Tiberghien, Open University Press, 2000.
- *MOSART (Misconceptions-Oriented Standards-Based Assessment Resources for Teachers)*, www.cfa.harvard.edu/smgphp/mosart/.

3 Energy

Energy: Descartes, Leibniz, Joule and Einstein - Kinetic and Potential, Heat, Chemical and Nuclear.

I was absent from school on the day we learnt about energy. I remember lining up outside the classroom and being told by my friends. I had no idea what energy was. For a long time, I felt I was trying to catch up. I thought I had missed a key piece of information.

That feeling gradually disappeared as I practiced answering energy questions, getting them right. It turns out that this is the best any of us can hope for with energy - no one really understands it.

I didn't realise this until long after I finished my physics studies when I read the *Feynman Lectures* - transcripts of lectures given by Richard Feynman:

> It is important to realize that in physics today, we have no knowledge of what energy "is." We do not have a picture that energy comes in little blobs of a definite amount. It is not that way. It is an abstract thing in that it does not tell us the mechanism or the reason for the various formulas.
>
> (Feynman 2010: section 4.1)

If Richard Feynman says we don't understand it, we don't understand it.

This is important. If you present energy as though it were something concrete, something understandable, you will leave students feeling lost and uncertain. We should share the secret: no one understands energy, but we can learn how to use it. And it is brilliantly useful.

A story is the best way to communicate and appreciate this secret knowledge.

A short history of five energies

I have written about five 'types' of energy because they tell the story of the development of an idea: *though we know that energy is conserved, we don't understand what energy is.*

People have always had an idea of energy. Anyone collecting or chopping firewood has a concept of energy. Anyone pulling a plough has a concept of mechanical work. Later, as medieval clockmakers developed systems of weights and springs, they must have understood that energy could be stored. So it is surprising that natural philosophers did not consider energy until the seventeenth century. This chapter begins with a short history of energy from that time, starting with kinetic energy and ending with $E = mc^2$.

Kinetic energy and potential energy: Descartes and Leibniz – 1644 and 1676

Once Copernicus had described the solar system, scientists began thinking about how the planets and moons kept in motion. What did they have that kept them going? Two conflicting ideas were developed: momentum and kinetic energy.

First, in 1644, Rene Descartes developed an idea he called *motion*, which he defined as the product of mass and speed (a version of momentum but without direction):

> It is obvious that when God first created the world, He not only moved its parts in various ways, but also simultaneously caused some of the parts to push others and to transfer their motion to these others. So in now maintaining the world by the same action and with the same laws with which He created it, He conserves motion; not always contained in the same parts of matter, but transferred from some parts to others depending on the ways in which they come in contact.

> (Descartes 2012: 62)

Descartes' idea was that God put a certain amount of motion into the universe when He created it, and the total of this motion has never since increased and never since decreased.

In Germany, Descartes' *motion* idea was developed by Leibniz, who realised that speed was the wrong measurement: it is velocity. We now call motion *momentum*, and the total amount of momentum in the universe is constant.

But Leibniz was not satisfied with motion. He was looking for the universe's spring, its animating force. He named his quantity of movement the *Vis Viva*, or the *living force*, which he calculated as the product of the mass and the velocity squared – in modern language: kinetic energy.

But the kinetic energy in the universe is not constant. It is not conserved. It is possible to create more and to destroy it, for example when a catapult launches a stone or when a ball is dropped into soft clay.

This un-conserved property of *Vis Viva* at first appears to be a disadvantage, but in fact it is this property which makes *Vis Viva*, or kinetic energy, so useful. *Vis Viva* can be transformed into something else.

To explain the transient nature of the *Vis Viva*, Leibniz invented another quantity. He called it *Vis Mortua*, or *the dead force*, which can be stored in compressed springs and in balls at the top of slopes. We now call it potential energy. *Vis Mortua* can be converted into *Vis Viva* and back again. Leibniz had invented mechanical energy.

There was instantly rivalry between supporters of Descartes and Leibniz. Newton believed that only Descartes' motion was required to explain the motion in the universe; his three laws make no use of energy at all. The majority of philosophers followed Newton. But the practical men of science, the mechanics and the engineers, knew that motion alone was not enough.

The academic case for kinetic energy, however, was made and won by Émilie du Châtelet, a French natural philosopher, mathematician and physicist. In 1740, she published a book, *Institutions de Physique (the Foundations of Physics)*, ostensibly written as a textbook for her son, but really an original work on natural philosophy. In it, she wrote a clear defence of *Vis Viva*. This led to a famous controversy and dispute with the well-known Jean-Jacques Dortous de Mairan, the secretary of the French Academy of Sciences. Du Châtelet countered every point made against her argument and de Mairan withdrew. *Vis Viva* was accepted.

Chemical energy and heat energy: James Joule – 1843

The next development in energy was developed by a man of business, James Joule, a successful brewer. His brewery used coal-powered steam engines. Joule wanted to know whether it was cheaper to buy coal or to use zinc batteries with electric motors. He needed to invent a way of measuring the output of the motor and engine to compare them.

His experiment was simple – to measure how much coal or zinc and acid were used to lift a weight a given height. As a businessman, he then worked out the cost. Coal was much cheaper.

Joule had measured the energy stored in the chemicals: coal and zinc and acid. Chemical energy. More than that, he had invented a unit for comparing energy: the *pound foot*. In other words, the amount of energy needed to lift a one-pound weight vertically upwards one foot. This was vital for his next discovery: heat energy.

Heat energy was a different matter. Heat already had an explanation. For most of scientific history, philosophers thought that heat was a type of substance. By 1667, philosophers called the substance of heat *phlogiston*. They believed that phlogiston was an element contained within all flammable substances and released when the substance burns. This idea was useful to scientists, but it couldn't survive the discovery of real elements, in particular the discovery of oxygen.

Phlogiston theory predicts that when a substance burns, it should lose mass as the phlogiston leaves the material. However, Lavoisier's discovery of oxygen showed that instead of phlogiston leaving during burning, oxygen joins the material and the mass increases. Lavoisier modified the phlogiston theory to account for this and renamed it the caloric theory. According to this theory, the quantity of caloric is constant throughout the universe, and it flows from warmer to colder bodies. But heat energy is not constant. It can be made.

Maverick scientists were discovering evidence that could not be explained by Lavoisier's key assumption: that the amount of caloric in the universe is constant. For example, boring a cannon produced a lot of heat. It could produce heat for as long as it was being bored. Either the metal had an infinite supply of caloric, or heat was being created.

Take a look at Figure 3.1; it shows a perfectly logical piece of equipment. The handle winds up the weight on the right, giving it gravitational potential energy. As the weight falls, gravitational potential energy is converted into kinetic energy in the paddles within the water.

The question is, *what happens to the kinetic energy when the motion stops?* You can see the thermometer on the left of the box. Joule showed that the gravitational potential energy from the weight is converted into heat energy in the water. Heat energy is not constant: this is new heat energy created from kinetic energy. Heat (caloric) is not conserved – energy is conserved.

Why was Joule ignored? Part of the problem was that if heat is an energy, it has to be carried by substances. The movement of atoms would be a suitable explanation, but the theory of atoms was not yet widely accepted (Joule was a student of John Dalton, the first modern scientist to describe the atom) and the idea of atoms moving was not established. The other reason was that Joule was not a professional scientist. It took many years before his work was recognised.

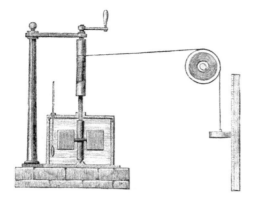

Figure 3.1 Joule's apparatus for measuring the conversion of gravitational potential into heat energy: by Unknown - *Harper's New Monthly Magazine*, No. 231, August, 1869. Public Domain, https://commons.wikimedia.org/w/index.php?curid=1527228

Nuclear energy: $E = mc^2$ - 1905

The genius of energy is that it can incorporate unexpected phenomena - heat does not look like kinetic energy or potential energy, but Joule found a way to incorporate it. Electricity, light and sound behave quite differently, but they too were incorporated.

But the story of nuclear energy is really unexpected. Albert Einstein showed that mass is a type of energy too. In an essay written 40 years after his discovery, Einstein wrote:

> Physicists accepted this principle up to a few decades ago. But it proved inadequate in the face of the special theory of relativity. It was therefore merged with the energy principle—just as, about 60 years before, the principle of the conservation of mechanical energy had been combined with the principle of the conservation of heat. We might say that the principle of the conservation of energy, having previously swallowed up that of the conservation of heat, now proceeded to swallow that of the conservation of mass— and holds the field alone.

(Einstein 2011: 14)

Einstein showed that the energy stored as mass can be calculated using the equation $E = mc^2$ (where E is the energy available, m is the mass and c is the speed of light - you get a lot of energy from a little bit of mass).

Physicists needed to understand the energy in atoms. Six years earlier Rutherford had discovered alpha and beta particles emitted from uranium atoms. These particles had kinetic energy - but where had the energy come from? Einstein's answer was that a very small amount of the uranium's mass was transformed into kinetic energy. In the language of GCSE exams today, *the energy in the nucleus' mass store is shifted into the particle's kinetic store*.

The equation is even more powerful than that. The beta particles emitted from Rutherford's uranium nuclei were not hidden inside the nucleus waiting to be released. They were brand new particles. Not only had mass been converted into kinetic energy of the beta particle, the mass of the beta particle had been created out of the energy of the nucleus. The subatomic world is fizzing with energy and particles: energy shifting from mass stores to kinetic and potential stores and back again.

The question is, what next? Energy seems to be future-proof. Whatever discoveries we make, energy will incorporate it.

Teaching energy

There are three ideas I want to share before I describe activities I use in teaching physics:

1   Don't try to explain energy: you can't explain energy. Definitions do not help. There is no lightbulb moment.
2   We learn how to solve energy problems through a gradual accretion of experience, of problems solved. The real route to knowing energy is by solving, and learning, all of the end-of-chapter problems from every textbook you can find.
3   The language of energy is important. Because energy is an abstract concept, we can't just show it to each other – we have to use words. Practice using the language precisely to help your students build accurate schemata.

Types of energy – stores and pathways

I was taught that there are nine *types* of energy. Over the last twenty years, this has fluctuated. My least favourite *type* of energy was solar – surely that's just heat and light?

The problem with *types* of energy is that some *types* behave very differently than others. It is more useful to categorise energy as *stores* and *pathways*.

In the UK, teachers from the Institute of Physics have taken this idea and proposed a better way of teaching energy. Kinetic energy, the potential energies, heat, chemical and nuclear energies are all described as *energy stores*, whereas electricity, sound and light are more helpfully described as *energy pathways*. (The Institute of Physics website has a full support programme for teaching energy – see Supporting Physics Teaching: Energy (En) (http://supportingphysicsteaching.net/EnHome.html).)

You can point to an energy store: a hot cup of tea; a petrol can; a cell; a photo of a bullet. The energy is associated with an object.

When we talk about the energy carried by light, sound or electricity, there isn't anything there. The energy is transient.

This has lost many generations of student. "I never got physics at school" is probably "I never got energy". No wonder.

Throughout this chapter, I have used the language proposed by the Institute of Physics (and used in GCSE assessments). It still doesn't come naturally to me, and, I'm sure, many older physics teachers. You, as a novice physics teacher, may have an advantage.

Misconceptions

Energy suffers from more misconceptions than any other area of physics – the word is so common and used in a haphazard way.

The answer is to practice the language in an accurate and consistent way.

Energy gets used up

The big idea of this topic is that energy is never lost. But that goes against everyday experience. The only way to get around this is to practice. Getting the language right helps:

- When I lift a box onto a shelf, energy is shifted from the chemical store of my arm to the gravitational store of the box.
- When a battery-powered car accelerates, energy is shifted from the chemical store of the battery to the kinetic store of the car.
- When I start to ride my bike, energy is shifted from the chemical store of my legs to the kinetic store of me and my bike.
- When a cup of tea cools down, energy is dissipated from the heat store in the tea to the heat store in the surroundings.

Avoid misleading language. A car does not *"use up"* energy stored in the petrol. The engine shifts energy from the chemical store of the petrol to the kinetic store of the car. But that doesn't explain why the petrol gets used up at a steady speed. As the car drives, energy is shifted from the kinetic store of the car to the heat store of the surroundings.

Work and energy

Another word which causes difficulty is *work*. Work sounds difficult: most learners would say that writing a page is more work than playing football for 10 minutes. But it isn't if you mean the amount of energy shifted from one store to another. Understanding the difference is fine, but unless you practice saying sentences like these, the meaning will never sink in.

- When I play football, I am doing work to shift energy from the chemical store in my muscles into kinetic and heat stores.
- When I solve a maths problem, the only work I am doing is shifting a small amount of energy from the chemical store in my brain into a heat store and a small amount of energy from the chemical store in my hand into the kinetic store in my pen.

Archetypal questions

Energy transformations

The basic energy task is to describe how energy shifts from one store to another when work is done. For example, an archetypal question is: *describe how energy is shifted in a lightbulb.* Practice these as often as you can - put them in the starter activities and quizzes and insist on the correct language.

To extend these questions, concentrate on shifting energy from gravitational stores to kinetic stores and back again. Examples include: balls rolling down slopes; projectiles fired vertically upwards; falling objects and pendulums. Once a student can accurately and reliably describe how the energy is shifted in a pendulum, you can be confident that she is ready for the next step.

Sankey diagrams

In Figure 3.2, the thickness of each arrow represents the quantity of energy.

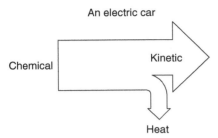

Figure 3.2 A Sankey diagram representing the shifting energy from a chemical store to heat and kinetic stores in an electric car

There is likely to be disagreement among physics teachers about which examples are best. I follow the Institute of Physics' advice – start and end with an energy store: chemical, kinetic, gravitational, nuclear, heat or elastic. Avoid electrical, light or sound. This may involve considering a system rather than an individual component – for example a toaster connected to a power station via the national grid (see Supporting Physics Teaching: Calculating Energy Everywhere – section 4 (http://supportingphysicsteaching.net/En04PN.html#PN2) for further examples).

An archetypal Sankey diagram question is: *if the electric car shifts 10,000J of energy from the chemical store in its battery, estimate how much energy is shifted in the kinetic store.*

This can be extended by including efficiency.

Calculating energy in stores

- $E_k = \frac{1}{2} mv^2$ (kinetic store)
- $E = mgh$ (gravitational store)
- $E = \frac{1}{2} kx^2$ (elastic store)
- $Q = mc\Delta T$ (heat store)
- $E = mc^2$ (nuclear store)

The first two of these equations are used in my archetypal question:

If a stationary body of mass 10 kg is dropped from a height of 5 m, calculate the quantity of energy shifted from the gravitational store to the kinetic store immediately before the object hits the ground.

I have modelled the answer below. Just as you should insist on the correct language for speaking and writing about energy, you should also insist on setting out the calculation clearly and consistently.

m = 10 kg
h = 5 m
g = 9.8 ms^{-2}

Energy in gravitational store at start:

E_g = mgh
    = 10 x 9.8 x 5
    = 490 J

Energy in kinetic store immediately before impact:

E_k = <u>490 J</u>

Note how I have written the initial data at the start, lined up the equals signs, written clear details and underlined the final answer.

There are several reasons for this attention to detail. First, your time in class is the student's most valuable resource. Don't waste that time trying to work out what the student means. As Lemov writes in *Teach Like a Champion 2.0*, "format matters" (Lemov 2015).

This format reduces cognitive load. The student does not have to think about the presentation, it is automatic.

Finally, mistakes are to be expected. This format makes mistakes easier for the student to spot.

Calculating electrical energy in a pathway

The next archetypal question is about energy pathways:

A battery-powered electrical device uses 3W. Calculate how much energy is shifted from the battery to the device in one minute.

P = 3 W
t = 1 min
   = 60 s

P = E/t
E = Pt
   = 3 x 60
   = <u>180 J</u>

Using strategies from cognitive psychology in lessons

Retrieval practice/spaced practice

Get your students to over-practice (practice until solving problems is automatic – not just on the day you teach it, but days, weeks and months later). When this knowledge and sentence type is automatic, your students will be able to use their working memories for more efficient learning and better problem solving.

Reading and writing

Following the Institute of Physics' advice on teaching energy stores and pathways is sensible. However, if students read about energy they will soon find texts using different language. I make these differences explicit.

Textbook language: In a filament lightbulb, electrical energy is transformed into light and heat energy.

Is equivalent to:

Institute of Physics language: In a filament lightbulb, energy arrives through an electrical pathway and is shifted to a light pathway – some energy is shifted into the bulb's heat store.

Reading is a powerful way of ensuring that your students are exposed to accurate language (teachers tend to simplify language to aid comprehension – you need to ensure that your students read and hear the precise language of energy).

Writing is a useful way of slowing down the language so that your students take their time in formulating sentences using the correct language. Drafting and rewriting allows students to produce precise language.

Reducing Cognitive Load

Modelling and completion problems

Modelling energy calculations reduces cognitive load and makes learning more efficient. It is very simple to reuse questions by changing the variables (for example 15 m instead of 10 m, 20 kg instead of 5 kg). You can change the context by changing the name of the object (for example a ball instead of a mass, or dropping from a tower instead of a shelf, cliff or balloon).

After modelling an answer, give your students half completed answers to complete. Eventually (over weeks and months), reduce the support until their answers are automatic. Insist on following your prescribed format.

Goal-free

Sankey diagrams are great for using the goal-free strategy (see Figures 3.3 and 3.4). Sankey diagrams are great for using the goal free strategy. I use solo-pair-share – one minute to write as much as they can independently (solo), then one minute to share with a partner (pair). While they are writing, go round and identify those who are struggling, but also those with helpful ideas to share. Ask those students to share their ideas with the class (share).

Example lesson plan

Outline plan (I have assumed a 60-minute lesson.)

This is a sample lesson which includes modelling a key calculation followed with completion problems. It is a Key Stage 4 lesson (14 to 16-year-olds).

I have assumed that your students are confident calculating kinetic and gravitational potential energy.

Time	Student activity	Teacher activity	My commentary
5 min	Starter activity: Retrieval practice (see Student activity 3.1).	Sort out any problems and then monitor completion. Discreetly encourage. Quietly sanction if necessary.	I've included retrieval practice followed by spaced practice and interleaving to support development of schemata.

(Depending on the time available, I would only do one of the following three activities – each learner should have the activity printed so time isn't lost copying.)

Time	Student activity	Teacher activity	My commentary
5 min	**Option A:** Tackling misconceptions using **refutation texts** (see Student activity 3.2).	Teacher monitors and identifies best answers.	I'd do this in three timed and very pacey stages. First they write for 90 seconds, you choose three students to read their answers (make brief notes on the board), then they have 60 seconds to rewrite. Repeat process for second text.
5min	**Option B:** *Solo-pair-share* **goal-free** task (see Student activity 3.3).	Teacher to monitor during the solo and pair stages, identifying the best ideas to bring out in share stage.	Again, I'd use a timer to keep the pace high – probably 60 seconds each for the solo and pair stages. I'd keep tight control of the share stage. Monitor whether individuals are updating their sheets.
5 min	**Option C:** *Solo-pair-share* **elaboration task** (see Student activity 3.4).	Teacher to monitor during the solo and pair stages, identifying the best ideas to bring out in share stage.	Again, I'd use a timer to keep the pace high – probably 60 seconds each for the solo and pair stages. I'd keep tight control of the share stage. Monitor whether individuals are updating their sheets.
25 min	Model answer and completion questions: Gravitational potential to kinetic energy Student activity 3.5).	Model one question and answer to students, then give remaining questions to students to complete.	I have modelled a sequence of similar questions, reducing the support in each question. Cognitive Load has been reduced, which will hopefully lead to more efficient learning of the technique, but you will need to repeat this exercise many times over months.
15 min	Reading: Conservation of energy.	In the texts I have recommended under 'My commentary', each sentence is nuanced. They are worth reading aloud, because the texts put carefully worded sentences and phrases in your students' mouths. They are also worth repeated reading, and then picking apart. For example, from Encyclopaedia Britannica: "Conservation of energy is a principle of physics according to which the energy of interacting bodies or particles in a closed system remains constant" (The Editors of Encyclopaedia Britannica 2017). The phrases 'interacting bodies' and 'closed system' are useful and precise, but you are unlikely to use them in classroom discourse.	Simple explanations of conservations of energy can be found across the web. For a few examples see: Mocomi 2015 (http://mocomi.com/law-of-conservation-of-energy/) BBC 2014 (http://www.bbc.co.uk/schools/gcsebitesize/science/add_ocr_pre_2011/explaining_motion/energychangesrev5.shtml) The Editors of Encyclopaedia Britannica 2017 (https://www.britannica.com/science/conservation-of-energy) (this entry is excellent, but you will need to give support to read it – it would benefit from a close read)

Time	Student activity	Teacher activity	My commentary
			Feynman et. al. 2010, Lecture 4 (http://www.feynmanlectures.caltech.edu/I_04.html) (this lecture should only be used if your group is confident: it is a short piece of text using metaphor to explain how energy always remains constant and how we know).
5 min	Review/exit-ticket.		

Student activity 3.1: Lesson starter quiz

1   A hot cup of tea stores which energy?

   a   Elastic
   b   Heat
   c   Kinetic.

2   Cars shift energy from:

   a   A kinetic store to a heat store
   b   A kinetic store to a heat store
   c   A chemical store to a kinetic store.

3   A projectile fired horizontally stores which energy?

   a   Kinetic
   b   Heat
   c   Sound.

4   Energy stored in an object's heat store can be shifted to another place by:

   a   Radiation, insulation and conduction
   b   Conduction, radiation and evaporation
   c   Convection, radiation and conduction.

5   A musical instrument shifts energy from:

   a   A sound pathway to kinetic store
   b   A chemical store in your body to a sound pathway
   c   An electrical pathway to a sound pathway.

6   Which sources of energy are renewable?

   a   Wind, solar, kinetic
   b   Geothermal, chemical, biofuels
   c   Solar, geothermal, biofuels.

7   A catapult shifts energy from:

   a   An elastic store in the catapult to a kinetic store in the projectile

   b   A kinetic store in the projectile to a heat store in the elastic

   c   A heat store in the elastic to a kinetic store in the projectile.

8   Wind turbines shift energy from:

   a   A heat store in the moving air

   b   A gravitational store in the moving air

   c   A kinetic store in the moving air.

Student activity 3.2: Tackling misconceptions – Refutation texts

These are easily adapted by switching the context. Instead of a petrol car, swap for an aeroplane or motorbike, etc. Instead of a hot drink, switch to a microwave meal, hot water bottle or radiator.

1   Some people think that when the energy in a car's chemical store (petrol) has run out, the energy no longer exists, but instead _____.

2   Some people think that when a hot drink cools down, the energy in its heat store is destroyed, but really _____.

3   People often say that their mobile phone has run out of charge, but really the energy _____.

Student activity 3.3: Goal-free activity sheet

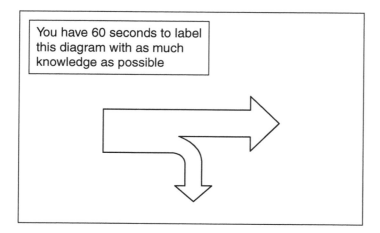

Figure 3.3 A blank Sankey diagram suitable for use as a goal-free activity

Student activity 3.4: Elaboration task

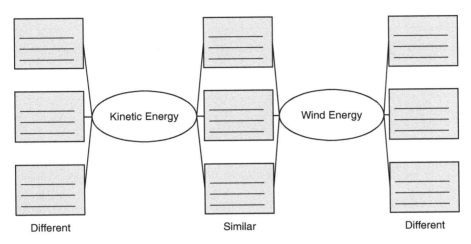

Figure 3.4 An activity for exploring similarities and differences between kinetic energy and wind energy

Student activity 3.5: Model answer and completion questions – Gravitational potential to kinetic energy

Completion question 3.5.1, with model answer: A 100 g ball is dropped 5 m from a balcony. Calculate the kinetic energy just before it hits the floor.

m = 100 g
   = 0.1 kg
g = 9.8 m/s²
h = 5 m

E_g before dropped:

E_g = mgh
   = 0.1 x 9.8 x 5
   = 19.6 J

E_k just before it hits the floor:

E_k = E_g
   = 19.6 J

Now it's your turn.
Completion question 3.5.2:

A 120 g ball is dropped 10m from a balcony. Calculate the kinetic energy just before it hits the floor.

m = ____g
   = ____kg
g = 9.8m/s^2
h = ____m

E_g before dropped:

E_g = mgh
   = ____ x 9.8 x ____
   = ____J

E_k just before it hits the floor:

$E_k = E_g$
   = ____J

Completion question 3.5.3

A 1.5 kg kitten jumps 2 m from a balcony. Calculate the kinetic energy just before it lands.

m = ____kg
g = ____m/s^2
h = ____m

E_g before dropped:

E_g = ____
   = ____ x ____ x ____
   = ____J

E_k just before it lands:

E_k = ____
   = ____J

Completion question 3.5.4

A 2 kg kitten jumps 1.5 m from a tree. Calculate the kinetic energy just before it lands.

m = ____
g = ____

h = ____

E_g before dropped:

E_g = ____

   = ____

   = ____

E_k just before it lands:

E_k = ____

   = ____

Completion question 3.5.5

A 600 g squirrel jumps 3 m from a tree. Calculate the kinetic energy just before it lands.

m = ____

   = ____

g = ____

h = ____

E_g before dropped:

E_g = ____

   = ____

   = ____

E_k just before it lands:

E_k = ____

   = ____

Completion question 3.5.6

A 400 g squirrel jumps 2.5 m from a tree. Calculate the kinetic energy just before it lands.

m = ____

   = ____

g = ____

h = ____

E_g before dropped:

E_g = ____

   = ____

   = ____

E_k just before it lands:

$$E_k = \underline{\hspace{1cm}}$$

$$= \underline{\hspace{1cm}}$$

Completion question 3.5.7

A 350 g squirrel jumps 2.5 m from a tree. Calculate the kinetic energy just before it lands.

_____

_____

_____

_____

E_g before dropped:

_____

_____

_____

E_k just before it lands:

_____

_____

Completion question 3.5.8

A 100 g book falls 1.5 m off a shelf. Calculate the kinetic energy just before it hits the ground.

_____

_____

_____

_____

E_g before dropped:

_____

_____

_____

E_k just before it lands:

_____

_____

Completion question 3.5.9

A 150 g book falls 1m off a shelf. Calculate the kinetic energy just before it hits the ground.

_____

_____

_____

_____

E_g before dropped:

_____

_____

E_k just before it lands:

_____

_____

Conclusion

When teaching energy, there are two points I really want to get across. First, energy isn't something we can understand - it is useful, but abstract, an accounting tool which helps us make calculations about the real world. The way to get comfortable with energy is to practise using it.

Second, the language of energy is important. Practice is the answer: speaking, reading and writing. The Institute of Physics recommend very specific language for energy (see Supporting Physics Teaching: Energy (En) (supportingphysicsteaching.net/EnHome.html)). Practical work, too, has a place, especially as a prompt for talking and writing.

Further reading – Energy

The following Institute of Physics (IoP) sites are useful:

- www.practicalphysics.org/energy.html has comprehensive instructions for teaching a wide variety of practical activities.
- http://supportingphysicsteaching.net/EsHome.html.
- www.talkphysics.org hosts a community of physics teachers which I find very helpful. Experienced colleagues answer questions and share advice, usually very quickly.

In addition, I recommend:

- *Children's Ideas in Science*, Driver, Guesne and Tiberghien, Open University Press, 2000.

- *MOSART (Misconceptions-Oriented Standards-Based Assessment Resources for Teachers)*, www.cfa.harvard.edu/smgphp/mosart/.
- *Physics for You*, Johnson, Oxford University Press, 2016.
- *The Feynman Lectures on Physics: Conservation of Energy*, Feynman, Leighton and Sands, California Institute of Technology 2010, www.feynmanlectures.caltech.edu/I_04.html.

4 Particles

All matter in the universe is made of very small particles.

(Harlen 2015: 15)

Introduction

In school, the first time the particle model of matter is introduced is when we teach solids, liquids and gases. We discuss the properties of each state and then we show a diagram of how the particles are arranged. The distance between 'liquids flow and take the shape of the container' and a diagram of circles in a box is huge. Stories can help to bridge the gap.

The first half of this chapter is a narrative of the development of our understanding of the particles which make up the universe. I have written it for the novice physics teacher, to provide a map and to begin developing your particles schema. This part of the chapter is intended to be read before you read and learn your medium-term plan and textbook.

The second half of the chapter is how I apply Cognitive Load Theory to the teaching of particles. I have started with misconceptions and then moved on to retrieval practice, archetypal questions, goal-free activities and comparison activities. I have written an example lesson plan for teaching solids, liquids and gases, which incorporates some of these ideas.

A history of particles

Not infinitely divisible - Atoms: The ultimate particles? Dalton - 1808

Matter, though divisible in an extreme degree, is nevertheless not infinitely divisible. That is, there must be some point beyond which we cannot go in the division of matter. I have chosen the word "atom" to signify these ultimate particles.

(Dalton 1810 quoted by Freund 1910: 288)

In 1808, the English schoolteacher, John Dalton, published a book: *A New System of Chemical Philosophy*. His 'new system' was revolutionary. He proposed that all matter was composed of tiny particles.

Before 1808, scientists and philosophers were struggling to understand the elements and the quantities in which they combine. Oxygen had been discovered in 1772, and by 1808 44 other elements had been named. Scientists knew that an element was a substance that could not be reduced further, but no one knew why. Dalton's idea, the atom, cut through the confusion.

His writing is clear and expressive. The extract 'On chemical synthesis' shows where he was right, but also where his model was inaccurate. He begins by explaining why a gas occupies more volume than the liquid from which it evaporated. He finishes the extract describing all of chemistry as the separating and re-joining of atoms: a brilliant explanation.

On chemical synthesis

For any gas, its particles are separated from each other to a much greater distance than in any other state. Each particle occupies the centre of a comparatively large sphere, and supports its dignity by keeping all the rest at a respectful distance.

When we consider the number of particles in a gas, it is like attempting to conceive the number of stars in the universe; we are confounded with the thought. But if we limit the idea, by taking a smaller volume of gas, we can believe that the number of particles must be finite; just as in a given space of the universe, the number of stars and planets cannot be infinite.

Chemistry goes no farther than separating particles one from another, and their rejoining. No new creation or destruction of matter is possible. We might as well attempt to introduce a new planet into the solar system, or to destroy one already in existence, as to create or destroy a particle of hydrogen. Any chemical change consists of separating particles that are joined, and joining those that were previously at a distance.

(Dalton 1808: 211–212)

Dalton's model has the atoms stationary and stacked like marbles, or piles of lead shot. He does not suggest that atoms vibrate or fly about. Instead, his atoms seem to puff out as the temperature increases:

Gases are constituted of particles possessing very diffuse atmospheres of heat, the capacity or bulk of the atmosphere being often one or two thousand times that of the particle in a liquid or solid form. Whatever therefore may be the shape or figure of the solid atom abstractedly, when surrounded by such an atmosphere it must be globular; but as all the globules in any small given volume are subject to the same pressure, they must be equal in bulk, and will therefore be arranged in horizontal strata.

(Dalton 1808: 145)

Dalton's book brought him worldwide fame, but it did not change his life: he remained a school teacher. This lovely quote by scientific historian E. J. Holmyard describes a visit by M. Pelletier, a French chemist, to Dalton:

What was the surprise of the Frenchman to find, on his arrival in Cottonopolis, that the whereabouts of Dalton could only be found after diligent search; and that, when at last he discovered the Manchester philosopher, he found him in a small room of a house in a back street, engaged looking over the shoulders of a small boy who was working his "cyphering" on a slate.

"Est-que j'ai l'honneur de m'addresser M. Dalton?" for he could hardly believe his eyes that this was the chemist of European fame, teaching a boy his first four rules.

"Yes," said the matter-of-fact Quaker. "Wilt thou sit down whilst I put this lad right about his arithmetic?"

(Holmyard 1896 quoted by Roscoe 1985: 54)

Dalton's atom was very small and indivisible: a perfectly hard little sphere. This idea didn't change for a century. The atom is a powerful model for explaining observations. But despite this success, the atom was at the centre of scientific debate for the next hundred years. The atom is useful, but is it real?

But atoms are not real. Or are they? Einstein - 1904

Do atoms exist, or is matter infinitely divisible?

(Maxwell 1873: 138)

Atoms are a powerful idea. Even before Dalton used the idea to explain ratios of elements in compounds, Daniel Bernoulli, the Dutch scientist, proposed a model of minute particles (corpuscles) to explain gas pressure. In 1738, he published *Hydrodynamica*, which included the first explanation of the kinetic theory of gases - the theory which explains pressure in gases through collisions of many, very small particles. In the following extract, *corpuscle* means *particle*.

Let the cavity contain very minute corpuscles, which are driven hither and thither with a very rapid motion; so that these corpuscles, when they strike against the piston and sustain it by their repeated impacts, form an elastic fluid which will expand of itself if the weight is removed or diminished.

(Bernoulli 1738 quoted by Datta 2005: 284)

Despite their power to explain phenomena, many scientists believed that atoms did not really exist. There was no proof: you cannot detect an atom.

The 19th century's great champion of atoms was Ludwig Boltzmann (1844–1906). He spent his career developing kinetic theory, a model of how atoms interact in great numbers, making very accurate predictions and explanations about the behaviour of materials and gases which could be tested by experiment. His contribution was the study of the random behaviour of atoms in gases. His model agreed so closely to measurements that he could not doubt their reality.

Boltzmann spent the final years of his life defending atoms. His great adversary, Ernst Mach, believed that atoms could only ever be a model, and should never be considered real: "The atom must remain a tool for representing phenomena, like the functions of mathematics" (Nola and Sankey 2001: 69).

During the final years of the 19th century in Vienna, philosophers and scientists battled for their ideas. Eventually, Boltzmann's view of atoms became the accepted model, but by then he had developed serious mental health problems. In 1904, proof was finally found that atoms are real. One year later, Boltzmann was dead.

Proof of the atom: Einstein and Brownian motion

1904 was an extraordinary year for physics. It is often referred to as *'the annus mirabilis'* – *the miraculous year*. Three papers were published by a then unknown physicist; each paper changed the world of physics. The physicist was Albert Einstein.

The papers solved three problems that had acted as barriers to progress in physics. All of them sound obscure, but they mark the start of the astonishing progress of 20th century physics. The first paper solved a problem about the emission of electrons from metal surfaces (the *photoelectric effect*). The solution involved light behaving as a particle rather than a simple wave. This is the paper that won Einstein the Nobel Prize for Physics.

The third paper he published is the most famous. In it he explains special relativity – how mass, length and time change when viewed from a moving viewpoint.

But it is the second paper, about pollen grains in water, that is important in this story. Its title didn't need to be catchy (*On the Motion of Small Particles Suspended in a Stationary Liquid, as Required by the Molecular Kinetic Theory of Heat*[1]). Its conclusion guaranteed it would be read:

> In this paper it will be shown that according to the molecular-kinetic theory of heat, bodies of microscopically-visible size suspended in a liquid will perform movements of such magnitude that they can be easily observed in a microscope, on account of the molecular motions of heat.
>
> (Einstein 1956)

The microscopically-visible objects were particles ejected from pollen grains. Einstein showed that the movement of these particles was caused by collisions with moving atoms.

This effect was known since 1827, when Robert Brown, a botanist who pioneered the use of microscopes, noticed that microscopic particles ejected from grains of pollen move randomly. No one understood what tiny engines could drive the particles. It was thought that the living particles had their own method of propulsion, until Brown showed the same motion in tiny particles of rock, smoke and other non-living materials. The movement of the particles was a mystery. It was named *Brownian motion*.

Seventy-eight years later, Einstein realised that many small collisions with tiny particles would cause the pollen to drift. His calculations showed that the collisions would combine to make the pollen jiggle. He had found the proof that atoms existed: atoms were the engines that moved the pollen.

It is impossible to observe the impact of a single atom colliding with a pollen grain, they happen too often (10^{21} per second!), but Einstein used mathematics to show that the random collisions of millions of atoms with a pollen grain will cause the pollen to drift to a new position which can be observed with a microscope.

This is as close to proof as physics gets. Atoms are real. The idea started as a model to explain observations. The model developed and the evidence accumulated until it was reasonable to conclude that the model was reality. Einstein had proven the existence of atoms.

Rays, beams and other phenomena – 1869 to 1899

Atoms were not the only topic for study. In the final years of the 19th century, physicists revealed a new world of mysterious, invisible beams, rays and particles. One discovery lead to another.

Cathode rays were the first to be discovered. In 1869, Johann Hittorf, a German physicist, discovered that he could cause a beam of electrical current to flow through a vacuum tube. A fluorescent material at the end of the tube glowed when the beam hit it. Hittorf discovered that the ray could be deflected by a magnet and would produce a shadow if an object were placed in the beam. Because the beam appeared at the cathode of the tube (the negative end), the beam was called a *cathode ray*.

In 1895, similar equipment was discovered to produce a new type of ray, the *x-ray*, which could travel through air, glass and many other substances, including human flesh. Wilhelm Roentgen discovered these rays while experimenting with cathode rays. He discovered that when a high-energy cathode ray struck a metal plate inside the vacuum tube, rays were given off that could travel across the room.

> If the electric discharge from a sufficiently exhausted Crookes tube, and if the tube is covered with a fairly close-fitting envelope of thin black card, it will be found that a paper screen placed near the apparatus and covered with barium platino-cyanide will become brightly luminous and fluorescent. It is immaterial whether the prepared side or the unprepared side is turned towards the apparatus. The fluorescence is still noticeable at a distance of two metres from the apparatus. It is easy to establish that the cause of the fluorescence proceeds from the discharge tube and from no other part of the electric circuit.

> The first remarkable feature about this phenomenon is that we have here an agent that can pass through a black card envelope which is impervious to the visible and ultra-violet rays of the sun or electric arc, and that this agent is capable of producing vivid fluorescence.

> Roentgen 1895

Just one year later, in 1896, Henri Becquerel was using naturally fluorescent minerals to study the properties of x-rays. He believed that uranium absorbed energy from the Sun and then re-emitted the energy as x-rays. He used photographic plates to detect the rays.

In February 1896, his experiment failed because there was not enough sunlight (it was cloudy). Nevertheless, he decided to expose the photographic plates anyway. The plates showed evidence of rays from the uranium despite the lack of sun.

Becquerel showed that his new rays could not be x-rays: x-rays cannot be deflected by a magnetic field. The deflection he demonstrated showed that the new radiation was charged, like the cathode ray. However, his new ray could travel through materials, more like x-rays. His deflections showed that there were positive rays, negative rays and neutral rays.

Pierre and Marie Curie took Becquerel's work further, examining the products of uranium and discovering the elements polonium and radium. Marie Curie named these rays *radioactivity*.

In 1899, a physicist from New Zealand, Ernest Rutherford (who gets most of the next section to himself), named two of these new rays: *alpha* (α) and *beta* (β). Alpha is positively charged and easily absorbed by air and thin sheets of metal. Beta is negatively charged and penetrates further through metals such as zinc. *Gamma* is neutral and takes relatively thick layers of lead to stop.

These rays and particles would be used to explore the internal structure of the atom. They were about to change everything that physicists thought they knew. The end of the 19th century for physicists was like arriving at an unexplored land.

Pieces of atoms – 1897 to 1899

Dalton's model of perfectly hard spherical atoms could not survive the discovery of the new particles and rays. Strangely, even before Einstein's proof that atoms were real, Dalton's model had to be modified: bits could get knocked off.

It is curious that physicists were already exploring the internal structure of the atom before the existence of atoms was proven. In 1897, British physicist J. J. Thomson published a paper describing how a new type of particle could be removed from an atom. He called these particles *corpuscles*: they were later renamed *electrons*.

Thomson knew about cathode rays, the currents that travel as beams through a vacuum. Opinions differed whether cathode rays were *aetherial*, like light, or *material*, made from particles. Thomson proved that the ray was a beam of negatively charged particles.

He used the following piece of equipment in Figure 4.1.

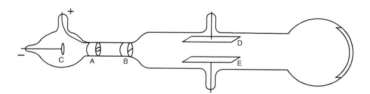

Figure 4.1 From *Philosophical Magazine*, **44**, 293 (1897)

A cathode ray is produced at C and accelerated through A and B. The plates D and E have a voltage applied to them. The cathode ray is deflected by the voltage. It is attracted to the positive plate, proving that the ray is negatively charged.

When Thomson applied a magnetic field across the beam, he could deflect the beam in the opposite direction (more evidence that the beam was negatively charged). But, by adjusting the magnetic field strength and the voltages across the plates at the same time, he was able to balance the two deflections. This allowed him to calculate the ratio of charge to mass.[2] This can only happen if the ray is made of particles.

In 1904, Thomson published a paper describing how these small negatively-charged particles might be arranged within an atom. He proposed a model for the atom with the main body of the atom as a positive charge with negative electrons embedded in it, ready to be knocked off. His model became known as the *Plum Pudding Model* of the atom:

We suppose that the atom consists of a number of corpuscles moving about in a sphere of uniform positive electrification: the problems we have to solve are (1) what would be the structure of such an atom, i.e. how would the corpuscles arrange themselves in the sphere; and (2) what properties would this structure confer upon the atom.

(Thomson 1904: 256)

So evidence proving the existence of electrons was found seven years before the evidence proving the existence of the atom. And just months before Einstein's Brownian motion paper was published, Thomson published his proposal of the internal structure of the atom. We knew about the structure of the atom before we were sure atoms were real. But even that model was short lived. Just seven years later, a totally new model of the atom was proposed: the nuclear model.

Disproof of the pudding: Rutherford's astonishing career - 1900 to 1921

This section is about the contribution made to physics between 1900 and 1921 by one man: Ernest Rutherford (Figure 4.2).

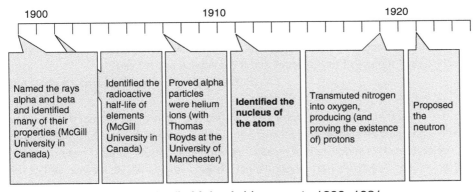

Ernest Rutherford's Major Achievements 1899–1921

Figure 4.2 Ernest Rutherford 1908

Ernest Rutherford was born and grew up in a small town in New Zealand, far from the academic world of physics. He was a very able student and in 1894 was awarded a scholarship enabling him to study under J. J. Thomson at the Cavendish Laboratory in Cambridge. This was the right place at the right time. From there, he went to McGill University in Canada, where his work on the properties of radiation earned him a Nobel Prize in Chemistry "for his investigations into the disintegration of the elements, and the chemistry of radioactive substances" ("The Nobel Prize in Chemistry 1908" 2018).

In 1907, he moved to Victoria University in Manchester and then finally back to the Cavendish Laboratory in Cambridge in 1919.

Rutherford's great success was in using alpha particles as keys to unlock the workings of the atom. He used them throughout his career in a series of elegant and important experiments, revealing the structure of the atom and identifying new particles. In 1899, Rutherford and Villard had named the alpha and beta rays (he named the gamma ray in 1903). In 1900, he published his findings on the radioactive decay of a product of thorium (radon), noting that the time taken for the activity to fall to half of its initial value was constant. He called this time the half-life period (later shortened to *half-life*).

In 1911, Rutherford made his greatest discovery: the nucleus of the atom. The experiment which involved hours of counting small flashes of light in a microscope (he didn't do this himself!), is described in Rutherford's own words below:

> I had observed the scattering of alpha-particles, and Dr Geiger in my laboratory had examined it in detail. He found, in thin pieces of heavy metal, that the scattering was usually small, of the order of one degree. One day, Geiger came to me and said, "Don't you think that young Marsden, whom I am training in radioactive methods, ought to begin a small research?" Now I had thought that, too, so I said, "Why not let him see if any alpha-particles can be scattered through a large angle?" I may tell you in confidence that I did not believe that they would be, since we knew the alpha-particle was a very fast, massive particle with a great deal of energy, and you could show that if the scattering was due to the accumulated effect of a number of small scatterings, the chance of alpha-particles being scattered backward was very small. Then I remember two or three days later Geiger coming to me in great excitement and saying, "We have been able to get some of the alpha-particles coming backward ...." It was quite the most incredible event that ever happened to me in my life. It was almost as incredible as if you fired a 15-inch shell at a piece of tissue paper and it came back and hit you.
>
> (Rutherford 1931 quoted by Cornford et al.: 68)

Rutherford realised that the recoil of alpha particles had two main implications. First, because most of the alpha particles went straight through the foil, most of the atom is incapable of causing a deflection. This suggests that most of the atom is empty, or at least has no charge. Second, the few alpha particles that rebound must 'strike' a very small, very dense part of the atom. In 1912, Rutherford named this small dense object the *nucleus*.

It is common in physics textbooks to see a third conclusion. Whatever caused the deflections must have a positive charge, because the positive alpha particles can only be repelled by a positive charge. But this is not correct. Attraction to a negative nucleus would cause a similar deflection. A positive nucleus was implied because negative electrons could be removed, suggesting the nucleus was positive.

So, the Plum Pudding Model of the atom was wrong. The atom has a nucleus. The positively charged pudding was replaced by a very small, very dense nucleus with a positive charge. The electrons were no longer embedded in the structure, but somehow existed within the main volume of the empty atom.

Rutherford's contributions to understanding the atom were not finished: there were still more discoveries for his alpha particles to make. Rutherford's next discovery was that the nucleus was not a single object (we say it is not a *fundamental particle*), but made of even smaller particles.

Physicists had speculated that it should be possible to change one nucleus to another by colliding them with high energy particles. In 1919, Rutherford was the first to knowingly *'split the atom'* (it had been happening, unrecognised, for as long as alpha particles had been used to bombard gases). As you can see in Figure 4.3, he bombarded a cylinder full of nitrogen gas with alpha particles.[3] Thin silver foil was used to absorb the remaining alpha particles. Beyond the foil, a fluorescent screen was used to observe whether any particles penetrated the silver foil. The flashes of light from the screen showed that some particles did penetrate (alpha particles would all be absorbed). These new penetrating particles were *protons*. Oxygen was found in the cylinder, confirming the conclusion.

$$^{14}_{7}N + ^{4}_{2}\alpha \rightarrow ^{16}_{8}O + ^{1}_{1}p$$

Figure 4.3 A diagram showing nitrogen absorbing alpha particles producing oxygen plus protons

The formula shows a modern notation representing the reaction. N is nitrogen, α is an alpha particle, O is oxygen and p is a proton. The top number (14) represents the mass of the nitrogen and the bottom number (7) represents the number of protons. You can see that the total mass on either side of the arrow does not change, nor does the number of protons.

Back in 1897, J. J. Thomson had found the mass to charge ratio for electrons using electric and magnetic fields. Rutherford used the same method to find the charge to mass ratio of alpha particles and protons. But something was wrong. The mass of the alpha particle was twice as big as it should be. Was there something else in the alpha particle?

Beta particle emission was also causing problems. When a nucleus emits a beta particle, it gains a proton. Where did this proton come from? In addition, beta particles have the same charge to mass ratio as electrons. Where did this electron come from?

An obvious solution was that there is another particle in the nucleus: a neutral particle composed of a proton and a neutron bonded together. In beta decay, the electron is released, leaving the proton behind. This solution explains why the alpha particle's mass is twice what it should be: it has extra neutral particles in it. In 1920, Rutherford published a paper proposing a hypothetical new particle. He named it the *neutron*.

The proton-electron version of the neutron helped explain beta decay, in which a nucleus ejects an electron and gains a proton. The idea was that the neutron simply split apart. The neutron was real, but took another 12 years to prove. However, the proton-electron model of the neutron was wrong.

Neutrons and war - 1932 to 1945

Neutrons were alpha particles' last great revelation. In 1932, Chadwick bombarded beryllium with alpha particles producing neutrons, as can be seen in Figure 4.4.

$$^{9}_{4}Be + ^{4}_{2}\alpha \rightarrow ^{12}_{6}C + ^{1}_{0}n$$

Figure 4.4 A diagram showing beryllium absorbing alpha particles producing carbon plus neutrons

He wasn't the first to do this experiment. In 1930, the German physicists Walther Bothe and Richard Becker were investigating a radiation that is emitted by beryllium when bombarded with alpha particles. They assumed it was gamma radiation.

In 1932, Irène and Frédéric Joliot-Curie investigated the same radiation. When the radiation hit paraffin wax, the wax emitted protons. However, they also interpreted the radiation as extremely high-energy gamma rays.

Rutherford and Chadwick did not believe that gamma had sufficient energy. Chadwick repeated the Joliot-Curie experiment with different materials instead of paraffin. He concluded that the alpha particles were causing the beryllium to emit neutrons. He published his short paper in *Nature* in February 1932.[4]

Rutherford's neutron was real. But his idea of a combined proton-electron model was not. The proof against it came through careful measurement. Binding the proton and electron requires energy. This energy can be measured in the mass of the neutron ($E = mc^2$). The mass of the electron and proton and the mass caused by the bond could be calculated. If the mass of the neutron were greater than this, then the neutron cannot be a composite. The mass was calculated in 1932 by Chadwick, proving that the neutron was not a composite.

So if the neutron was not a joined proton-electron, what was it? The explanation was found in 1934 by the Italian physicist Enrico Fermi. However, his solution was so unexpected that *Nature*, the respected journal, refused to publish it. Fermi showed that a neutron could decay into a proton, releasing a beta particle plus a hypothetical particle, the *neutrino*[5]. Fermi was the first to show how particles could be created or destroyed. His ideas lie at the heart of all modern nuclear physics.

Fermi was so discouraged by *Nature's* refusal to publish that he took a break from theoretical physics. Instead he began experimenting with slow beams of neutrons, leading to the processes making nuclear fission possible.

In 1938, Fermi left fascist Italy with his Jewish wife. He went to the US and made valuable contributions to the Manhattan Project, resulting in the first atomic bomb.

The Manhattan Project gathered together physicists from across Europe and the US; it was the greatest collaboration of physicists. Their goal was to develop an atomic bomb before the Nazis did. The war in Europe had ended before the bomb was ready, but the war in the Pacific continued. Two atomic bombs were dropped on Japan in 1945.

After the war, European physicists were determined to develop a peaceful project to bring the physicists of Europe together. The collaboration was CERN, the European Organization for Nuclear Research, which remains one of the world's leading particle physics research establishments.

Teaching particles

The story of particle physics is one of the great achievements of physics – and the story is ongoing. It is a story of theoretical models years ahead of proof (kinetic theory, the atom, the neutrino), punctuated by unexpected discoveries (cathode rays, x-rays, alpha scattering).

Your students need to learn the models and how to apply them. For most of us, this is an act of faith. I find the history of particles helpful, because it shows the struggle for understanding, which your students may empathise with!

Misconceptions

We are born with some physics misconceptions. Our intuitive, pre-loaded misconceptions about how objects behave isn't a problem early on when we learn about particles. Classical particles behave sensibly, like marbles (sometimes sticky marbles). But quantum particles trigger our misconceptions like nothing else in physics: particles can be in many places at once; they can appear and disappear; they can be waves as well as particles. We are not born to understand quantum mechanics.

But learners can develop misconceptions on the way. They usually first learn about particles when learning the states of matter. Common misunderstandings, which can be persistent, include:

- Confusion between melting and dissolving.
- Particles disappear when evaporating or dissolving.
- Evaporation only happens when a liquid boils.
- Gases have no mass.

Beware these misconceptions – it is worth assessing for them even years later. Refutation texts are useful. For example, give students these sentence starters:

Many people think that gas has no mass, but _____.
One difference between melting and dissolving is _____.
Many people think that water molecules vanish when water evaporates, but _____.

Once the foundations are strong, the building can begin.

Archetypal questions

There are two questions which lie at the heart of particle physics:

- How are the particles arranged in a solid, liquid and gas?
- How do gas particles exert pressure?

The first becomes important at Key Stage 2 (though they may have learnt it at primary school). The second becomes important later. These questions need to be practiced many times to be learnt. I have given various elaborations on these questions. I start by modelling the answers. Then give *completion problems*, which reduce Cognitive Load by asking the learner to complete one key part of the solution at a time. The *goal-free* strategy works well with both of these questions – show a text-free image and ask students to annotate it with as much knowledge as possible (then can then share with a partner before you identify two or three students to share key ideas with the class). There is nothing wrong with repeating these activities many times – *overlearning* is an effective way of ensuring what's been learnt stays learnt.

I have also identified questions about radioactivity which are simple to answer, but key to developing the schema.

The particle model

Archetypal task: *Draw circles to show the arrangement of particles in a solid, liquid and gas.*

ELABORATIONS

Students need to be taught good answers to these questions before you quiz them. Insist on high quality answers – they can memorise sentences.

- Describe how the particles move.
- Describe how the particles are arranged.
- Describe how the particles interact.
- Explain what a particle represents (i.e. an atom or molecule).
- Use the particle model to explain why solubility increases with temperature.

COMMON MISTAKES

- Particles expanding as you heat them. Train your students to use uniform circles.
- Students draw small circles, which take too long to complete. Train them to use penny-sized circles.
- Students may not realise what a particle represents.

Kinetic theory

Archetypal question: *Explain how gas particles exert pressure.*

KEY KNOWLEDGE

1 Gas particles collide with the walls of a container.
2 When the particles collide with a wall, they change direction.
3 The wall of the balloon provides the force to change the particles' direction.
4 Newton's third law explains that when the wall exerts a force on the gas particle, the gas particle exerts an equal and opposite force on the wall.

- Avogadro's principle explains that under the same pressure and temperature conditions, equal volumes of all gases contain the same number of particles.
- The gas laws:
    - Boyle's Law: For a fixed amount of gas kept at a constant temperature, pressure and volume are inversely proportional.
    - Charles' Law: When the pressure on a sample of gas is constant, the temperature (in Kelvin) and the volume will be directly proportional.
    - Gay-Lussac's Law: For a fixed mass of gas, at a constant volume, the pressure is directly proportional to the temperature in Kelvin.

ELABORATIONS

- Explain how the air particles outside a balloon affect the size of the balloon.
- Describe what happens to the size of a balloon as the air pressure outside the balloon decreases.
- Explain why the volume of a balloon increases as the air pressure outside the balloon decreases.

- You have two identical balloons, one filled with carbon dioxide and one filled with helium. What do you know about the speed of the gas particles? (Helium has less mass, so the particles must be travelling faster. A particle of CO_2 is 11 times more massive than a helium particle. The velocity of helium particles is $\sqrt{11}$ times greater.)
- What happens to the volume of a balloon as the temperature of the gas inside increases?
- Explain why a bubble of gas released by a diver 30m below the surface of the water will expand as it rises (ignore any effects of temperature changing).

Radioactivity and the atom

I haven't chosen an archetypal question for radiation, but there is a family of tasks which are asked repeatedly in examinations and are vital for understanding radioactivity:

- Make sure your students can identify the components of an atom.
- Make sure your students can balance nuclear equations.
- Make sure they can interpret half-life graphs.
- Make sure they memorise the nature and properties of alpha, beta and gamma radiation.
- Make sure they know the definitions of fission and fusion.

They should also be able to write longer explanations of the following:

- How Rutherford's gold-leaf alpha scattering experiment is carried out and what it showed.
- How gamma sources can be used as medical tracers.
- How a nuclear power station uses heat to generate electricity.

To do well in exams, your students need to know how to answer these questions, so they should practice them both during the topic and during subsequent topics (weekly *spaced retrieval* practice is an effective use of lesson time or homework).

Useful strategies from cognitive psychology in lessons

Retrieval practice – Quizzes

There are plenty of quizzes online (I use BBC *Bitesize*). There is nothing wrong with multiple-choice as long as the learner has to recall previous knowledge.

It doesn't hurt to repeat the same quiz – you are aiming for 100% correct – as overlearning is helpful. Swap the order around and paste in questions from other topics to make use of spaced retrieval practice.

Goal-free

The images of particles in solids, liquids and gases work well with a goal-free exercise. Show your students an unlabelled image to annotate with as much knowledge as they can. I use a timer to keep it fast. They can then compare with a partner, adding to their annotations.

While they are doing that, I usually assess by looking over shoulders. I pick two or three useful ones and ask those students to read their answers out.

Goal-free also works well with older learners. Try: kinetic theory of gases; dot-and-cross diagrams; Rutherford's gold-leaf alpha scattering; half-life decay curves; and Feynman diagrams. Using text-free diagrams from exam questions is also useful, as per Figure 4.5.

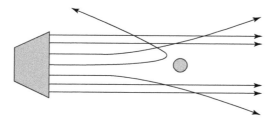

Figure 4.5 A goal-free activity with a diagram of Rutherford scattering

Similar/different

Use the similar/different activity to compare two states of matter or melting and dissolving. I've filled in an example in Table 4.1.

Table 4.1   Similar/different activity example

Different		Similar		Different
Fixed position		Touching		Free to move about
Vibrating	**Particles in a solid**	Bonded together	**Particles in a liquid**	Move faster when hotter
Bigger vibrations when hotter		The particles don't change size when heated		Weakly bonded to other particles

Reading

Once students have developed a decent particle schema, you can introduce reading. In class, I ask students to read aloud (I use the strategy called *Control-the-Game* from *Teach Like a Champion 2.0* or *Reading Reconsidered*, both by Lemov).

The key strategy to make a read aloud activity effective is to swap readers quickly and unpredictably – each student only reads a sentence or two before you ask another to take over. You don't need to praise or comment, just call the next name to keep the text flowing. This takes a little practice to make fluent, but usually only two or three goes.

Choose a text which reviews knowledge they already have. They should already know most of what they are reading – it is more effective as a learning activity when students are already familiar with most or all of the content. The text can be short, so you can do it often.

Independent reading is useful, but you don't need to use valuable class time for that. Set it as homework, but make it accountable – comprehension questions are fine, but make it so they can't answer without reading the text. Asking them to paraphrase a paragraph works well, as does asking them to set their own questions based on the text.

Writing

Exams often ask students to write longer answers for kinetic theory and radioactivity. Earlier in this chapter I identified the following archetypal questions: *how Rutherford's gold-leaf alpha scattering experiment is carried out and what it shows; how gamma sources can be used as medical tracers;* and *how a nuclear power station uses heat to generate electricity.*

It is efficient for students to learn these texts by heart.

To reduce the Cognitive Load of writing, you can provide structure (for example bullet points), key vocabulary or sentence starters. Provide time for improving answers – I ask students to miss a line between each to provide space to correct or improve. Aim to reduce support over time.

Practical particles

There are several key demonstrations and practicals your students should learn at high school. For kinetic theory, these include:

- The thermal expansion of a metal
- Boyle's Law
- Physical models for kinetic theory (balls in containers).

Instructions for these can all be found at the Institute of Physics' *Practical Physics* website (practicalphysics.org).

Nuclear radiation demonstrations are straightforward, safe and reliable, but you need to do them correctly. A demonstration of alpha, beta and gamma radiation and their absorption can be quick and effective. You may choose to do it twice: once at the start of the teaching sequence, before you teach the models, and again when your students have demonstrated a clear understanding of the models.

Instructions for using nuclear radiation models can also be found at the Institute of Physics' *Practical Physics* website (practicalphysics.org) and elsewhere. Subject organisations such as the Association for Science Education (ASE) and the Institute of Physics (IoP) have guidance for safe and effective practical work.

Example lesson plan

Outline plan (I have assumed a 60-minute lesson.)

This is a sample lesson which includes a short practical activity. It is a Key Stage 3 lesson (for 11 to 13-year-olds) about changes of state.

I have assumed that the students have already had one or more lessons on the particle model and are familiar, though perhaps not fully confident.

Time	Student activity	Teacher activity	My commentary
5 min	Starter activity: Retrieval practice (see Student activity 4.1).	Sort out any problems and then monitor completion. Discreetly encourage. Quietly sanction if necessary.	*I've included retrieval practice followed by spaced practice and interleaving to support development of schemata.*
	(Depending on the time available, I would only do one of the following three activities – each learner should have the activity printed so time isn't wasted copying.)		
5 min	**Option A:** Tackling misconceptions using **refutation texts** (see Student activity 4.2).	Teacher monitors and identifies best answers.	*I'd do this in three timed and very pacey stages. First they write for 90 seconds, you choose three students to read their answers (make brief notes on the board), then they have 60 seconds to rewrite. Repeat process for second text.*
5 min	**Option B:** *Solo-pair-share* **goal-free** task (see Student activity 4.3).	Teacher to monitor during the solo and pair stages, identifying the best ideas to bring out in share stage.	*Again, I'd use a timer to keep the pace high – probably 60 seconds each for the solo and pair stages. I'd keep tight control of the share stage. Monitor whether individuals are updating their sheets.*
5 min	**Option C:** *Solo-pair-share* **elaboration task** (see Student activity 4.4).	Teacher to monitor during the solo and pair stages, identifying the best ideas to bring out in share stage.	*Again, I'd use a timer to keep the pace high – probably 60 seconds each for the solo and pair stages. I'd keep tight control of the share stage. Monitor whether individuals are updating their sheets.*
25 min	Direct instruction using a model and demonstration of Boyle's Law (see Student activity 4.5).	Refer to Student activity 4.5 for instructions.	
10 min	Reading: A biography of Robert Boyle.	Prepare the reading as per the reference under 'My commentary'.	The biography of Robert Boyle (see The Doc 2015 (https://www.famousscientists.org/robert-boyle/)) - is long, so only read two or three paragraphs. The rest is great for retrieval practice, so I might set it as homework with the instruction to describe three discoveries Boyle made with his vacuum pump.
10 min	Review/exit-ticket.	Remove the model explanation of Boyle's Law from the board and replace it with scaffolding for students to write their own account: 1. State Boyle's Law. 2. Explain how collisions cause pressure. 3. Explain why reducing the volume increases the pressure.	By writing, students are practicing using the main idea from the lesson. You have scaffolded the answer by using questions, reducing the Cognitive Load. If they need to do this in two stages (as answers first then as a paragraph), that is fine. If your students are struggling with answering the questions, you could do each question as a write/rewrite (see *Reading Reconsidered*, Lemov et. al. 2016), before they write the full version.

Student activity 4.1: Starter activity sheet

1   In which state do the particles vibrate in a fixed position?

    a   Solid

    b   Liquid

    c   Gas.

2   In which two states are the particles randomly arranged?

    a   Liquid and solid
    b   Liquid and gas
    c   Gas and solid.

3   Which state can be compressed easily?

    a   Solid
    b   Liquid
    c   Gas.

4   Which state cannot flow from place to place?

    a   Solid
    b   Liquid
    c   Gas.

5   Particles in which state are not bonded together?

    a   Solid
    b   Liquid
    c   Gas.

6   Particles in which state are arranged in a regular pattern and are held together by bonds?

    a   Solid
    b   Liquid
    c   Gas.

7   An object at rest on a shelf has:

    a   No energy
    b   Gravitational potential energy
    c   Kinetic energy.

8   Batteries are stores of:

    a   Chemical energy
    b   Electrical energy
    c   Kinetic energy.

9   What happens to an atom if it gains an electron?

    a   It becomes positive
    b   It becomes negative
    c   It stays neutral.

10   What charge are electrons?

    a   Neutral
    b   Positive
    c   Negative.

Student activity 4.2: Tackling misconceptions – Refutation texts

I find refutation texts useful in overcoming learnt misconceptions in particle physics. Sentence starters help, for example:

1   Many people think that there are gaps between the particles in a liquid, but really _____.

2   Many people think that particles in a gas are bigger than the particles in a solid, but really _____.

Student activity 4.3: Goal-free activity sheet

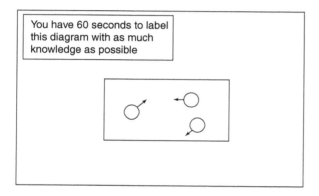

Figure 4.6 A goal-free activity with a diagram of particles in a gas

Student activity 4.4: Elaboration task

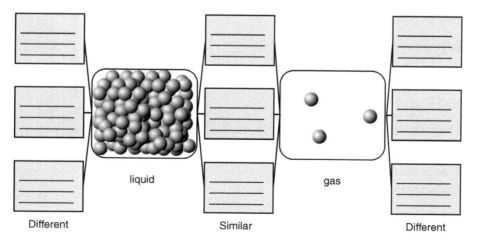

Figure 4.7 An activity for exploring similarities and differences between liquid particles and gas particles

Student activity 4.5: Direct instruction using a model and demonstration of Boyle's Law

Demonstrate a gas model using a tray of marbles (see Institute of Physics 2005). A visualiser would be very useful for this.

Model writing this sentence:
Boyle's law states that as the volume decreases, the pressure increases.

The students complete this sentence:
As the volume increases _____.

Use the gas model to explain that when you decrease the volume (i.e. decrease the available area of the tray), there are more collisions per second with the wall. This means the pressure increases.

You then model this text:
Gas exerts a pressure on the wall of a container by colliding with the walls. As the volume decreases, the number of collisions per second increases, increasing the pressure. This is called Boyle's Law.

Now show students the Boyle's Law apparatus and carry out the experiment as a demonstration. If you want to record the results, use a spreadsheet so that the graph can be shown quickly.

Note: If you want the students to experience the model for themselves, use two or three sets at the side of your class for students to try in pairs during the exit-ticket stage. Each pair only need one minute.

Conclusion

Teaching particles is about teaching models. The practical work that you can do is limited and it is challenging for students to link the evidence to the model. You have to show them how. Because your students have little first-hand experience, they will need lots of practice to become familiar with the problems and the solutions. In other words, you can't practice too much.

Notes

1 For a full translation into English, see Einstein (1956) (http://users.physik.fu-berlin.de/~kleinert/files/eins_brownian.pdf).
2 For an A-level demonstration of this calculation, see DrPhysicsA (2012) (https://www.youtube.com/watch?v=qd4AkHAe7mE).
3 For a good Rutherford foil and transmutation experiment explanation, see Owens (2009) (https://www.youtube.com/watch?v=uOHMYLSUzTU).
4 See Chadwick's original paper (1932) (http://web.mit.edu/22.54/resources/Chadwick.pdf).
5 For a history of the neutrino, see Casper (http://www.ps.uci.edu/~superk/neutrino.html).

Further reading - Particles

The following Institute of Physics (IoP) sites are useful:

- www.practicalphyiscs.org has comprehensive instructions for teaching a wide variety of practical activities.
- http://supportingphysicsteaching.net/RaHome.html.
- www.talkphysics.org. They host a community of physics teachers, which I find very helpful, where experienced colleagues answer questions and share advice, usually very quickly.

For misconceptions, the following are helpful:

- *Children's Ideas in Science*, Driver, Guesne and Tiberghien, Open University Press, 2000.
- *MOSART (Misconceptions-Oriented Standards-Based Assessment Resources for Teachers)*, www.cfa.harvard.edu/smgphp/mosart/.

5 The universe

Orreries and telescopes: physical orreries; mathematical orreries; Galileo's Telescope – 1609; Newton's Telescope – 1687; Edwin Hubble; Jansky's Merry-Go-Round.

Introduction

The first thing you need to know about space is that the Sun seems to move across the sky. You know that this illusion is a result of the Earth's rotation.

The second thing to know is that the stars also appear to revolve around the Earth, for the same reason. But you might not know that the stars revolve around the Earth faster than the Sun – four minutes faster each day.

The third thing to know is that while the stars do not move relative to each other, the planets move slowly through this pattern of stars. You can see this with your own eyes. If you look at the position of Venus, Mars, Jupiter or Saturn over several nights, you will notice its position change relative to the stars. The planets move in relation to the stars.

To understand these facts, you need a model. The first section of this narrative is about the development of this model – an orrery.

The second section is how the telescope was used to develop the model of the universe.

The final part of this narrative is how our understanding of space itself has changed since Copernicus published his heliocentric (Sun-centred) universe.

The orrery

Painted on a ceiling in the Lascaux caves in France is a pattern of stars – the Pleiades (the *seven sisters*). It is a beautiful star cluster, beautiful enough in the ice-age sky to record forever on the walls of a cave.

The cave is a planetarium without planets. A theatre of stars. A backdrop. The ice-age artists, the first astronomers, knew that the brightest dancers in the sky, the Sun, the Moon and the planets, move across the backdrop, moving through the patterns of stars, following their own paths across the stage.

Understanding the dance of the planets has been one of humanity's great intellectual stories. We now understand the motion of the planets so well that we can launch a space-probe from Earth, fly past Jupiter and on to Pluto, arriving, but not stopping, nine and a half years later. We have learnt to understand the machinery of our solar system by making mechanisms of our own: from ice-age paintings to machines that spin the planets in planetariums to models encoded in software.

The earliest mechanical universe was found deep in the Aegean Sea amongst the treasures of a Roman shipwreck. Sponge divers found a 2,000 year old mechanism able to predict the path of the Sun and Moon through the stars. This cogged machine has the Earth at the centre of the universe – it is geocentric – and it is accurate.

Machines like these are called *orreries*, mechanisms able to predict the journeys of the planets through the field of stars. They are mechanical computers. Every orrery holds its maker's ideas about the heavens in its gears. Through orreries we can trace the development of our understanding.

Simple geocentric orreries showing just the Sun, Moon and stars work well. The Sun appears to orbit the Earth in an approximately circular orbit, as does the Moon. The problems begin with the planets.

By holding the Earth still at the centre of the model, the planets do not appear to move in elegant ellipses or circles; they loop and dance. Sometimes they appear to move backwards among the stars, sometimes forwards. It is like watching a car from a moving train.

Orreries need clever additional mechanisms to keep track: rotating arms on the end of other rotating arms to create the looping patterns. The mechanisms became sophisticated enough to make accurate predictions. We didn't need a better model.

But we got a better model. It was a model to explain this motion – a model which places the Sun at the centre of the universe, a heliocentric model.

To move the Sun to the centre of the model was Copernicus' revolutionary idea. It was a dangerous idea. Men were tortured, imprisoned and killed for promoting it. Copernicus' great book is called *De Revolutionibus* – he meant revolutions in the sense of orbits, but we can't read the title without thinking about a scientific revolution. Copernicus got away with it: he died of old age, holding in his hands, for the first time, a copy of his newly printed book.

There are three common misconceptions about Copernicus' model that I want to challenge. The first misconception is that his model was simpler than the geocentric model: it wasn't. Copernicus only considered circular orbits, which resulted in inaccurate results. (He had to add rotating arms to his model, too.)

The second misconception is that Copernicus' model was more accurate than the geocentric model: again, it wasn't. The geocentric model was so sophisticated and refined that it was more accurate.

The third misconception was that Copernicus didn't believe his model was a true representation of reality. That isn't true either – he believed the planets orbited the Sun.

This third misconception was introduced by the printer in Nuremberg. A local scholar wrote the introduction which stated that the heliocentric model is useful as a tool, but should not be accepted as truth:

> Let no one expect anything certain from astronomy, which cannot furnish it, lest he accept as the truth ideas conceived for another purpose, and depart this study a greater fool than when he entered.

(Osiander 1543: ii)

Copernicus' great champion, Rheticus, was so outraged by this addition that he threatened to "rough up the fellow so violently that in future he would mind his own business" (MacLean 2008).

But the martyrs of physics came later. Giordano Bruno was burnt to death. Galileo Galilei was tortured and imprisoned.

Mathematical orreries – Orreries of the mind

While some of the early orreries were used as calculation tools, most model universes were used to explain rather than calculate. In around 150AD, the great astronomer Ptolemy wrote a book we now call the *Almagest* (the *Great Treatise*).

He recorded and explained how to calculate the motion of the planets with the Earth at the centre of the model – a geocentric model. Ptolemy's model was very accurate. When Copernicus developed his heliocentric model with the Sun at the centre of the universe 1,200 years later, his model was less accurate.

This is because Copernicus used circular orbits for the planets.

Copernicus' concept of the universe, with the Sun at the centre and circular orbits, is what we think of when we imagine an orrery. But the planets do not orbit the sun in circles, they move in ellipses.

Copernicus' model was not successful for calculation, but it triggered revolutionary thinking by Kepler, Galileo and Newton.

Kepler

Johannes Kepler made Copernicus' universe work. He had extremely accurate data to work from and he trusted it. The data was collected by his mentor, the Danish astronomer Tycho Brahe. Brahe was the last of the great astronomers to work before the telescope was invented. His observatory consisted of huge sextants and quadrants, capable of measuring the angles of the stars and planets with great precision and accuracy.

Kepler used this data to develop his three laws – the laws which made the heliocentric model accurate:

1   The orbit of a planet is an ellipse with the Sun at one of the two foci (ellipses have 2 'centres' called foci.)
2   A radius, the line joining a planet to the Sun, sweeps out equal areas during equal intervals of time. When the planet is closer to the Sun, it travels more quickly, so although the radius is smaller, the path is longer. When it is farther from the Sun, the radius is larger but the velocity is smaller. The result is that in one day, the 'slice of cake' that the planet moves has the same area regardless of where it is in the orbit. This is demonstrated in Figure 5.1.
3   The square of the orbital period of a planet is proportional to the cube of the semi-major axis of its orbit (half of the distance across the widest point of the ellipse).

Without these three laws, Newton could not have written his Law of Universal Gravitation. Kepler's improvements to Copernicus' model allowed Newton to unify astronomy with physics – one of the greatest achievements of science.

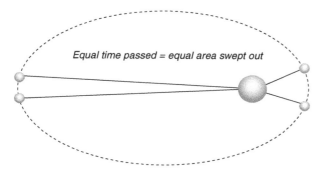

Figure 5.1 Diagram showing Kepler's 2nd Law where an orbiting body sweeps out an equal
area in equal time

The telescope - 1608

Seeing faraway things as though nearby.

(Metius quoted by Helden and van Gent 2010: 15)

You can make a telescope with two lenses. Try it. Take a fat, short focal-length lens and hold
it in front of your eye. Then take a slender, long focal-length lens and look through it at arm's
length. Align them both with a distant object. As you bring the slender lens closer, the image
will become clear - you have a telescope without the tube. It is said that two children discovered this in an optician's workshop in Holland in 1608.

Sometimes in science, the same discovery is made simultaneously by several people. The
telescope is an example of this. In the early 1600s craftsmen in Holland were making high
quality lenses. In 1608, three lens-makers invented the telescope independently.

Galileo's telescope - 1609

The Dutch telescopes could magnify an object three times larger than real life. One year after
its invention, Galileo made his own telescope which could magnify 20 times. He pointed it at
the Moon.

Galileo's telescope was so much better than anyone else's, he could make observations
that no one was able to verify. He made good use of his head-start.

From the 30th of November to the 19th of December 1609, he observed the Moon. He
saw craters, mountains and solidified lava lakes. This wasn't what you were supposed to
see. The heavens were supposed to be perfect, without blemish. The Moon was a jag-
ged bomb-site.

In January 1610, he discovered the moons of Jupiter, studying their motion from the 7th
of January to the 15th of January. He realised they were like a tiny Copernican solar system.

Galileo was relentless - he knew what to look at and how to make sense of what he saw.

In March 1610, Galileo published *Sidereus Nuncius* (the *Starry Messenger*).

He observed the phases of Venus in December 1610. The phases showed Venus was orbiting the Sun.

If Galileo had been more of a diplomat, he might have had an easier life. In 1610, he got
the politics right when he named Jupiter's moons the *Medicean stars*, after the Pope's family

name. Twenty-two years later, he got the politics very wrong when he published the *Dialogue Concerning the Two Chief World Systems*.

It was a dialogue between two men, one rational and precise (and right), the other vain and foolish. The rational character described Galileo's views. The foolish character (named Simplicio as additional insult) was given the words of a Jesuit priest, Father Christopher Schneider, who argued for Ptolemy's model of the heavens. This did not go down well.

Even worse, Galileo then put into Simplicio's mouth an argument made by Pope Urban VIII a decade earlier. In 1633, Galileo was found to be "vehemently suspect of heresy" (Shea and Artigas 2003: 194).

The Papal Inquisition interrogated Galileo. He confessed that the Copernican heliocentric model was wrong:

> After having been judicially instructed with injunction by the Holy Office to abandon completely the false opinion that the sun is the center of the world and does not move and the earth is not the center of the world and moves, and not to hold, defend, or reach this false doctrine in any way whatever, orally or in writing.
>
> (Galileo 1633)

There is a legend which says that Galileo then whispered "... and yet it moves".

He was allowed to live under house arrest under the condition that he never point his telescope at the night's sky again.

So he finally turned his telescope towards the Sun, gradually blinding himself. He saw dark spots slowly travelling across the surface. The Sun too, was irregular, blemished, imperfect.

Newton's telescope – 1687

It was a telescope that cemented Newton's reputation. In 1668 (19 years before his great work, the *Principia*, was published in 1687), Newton successfully replaced the slender objective lens with a curved mirror. It wasn't his idea, but no one had successfully made one before. We don't usually think of Newton as a practical craftsman, but his telescope proved that he was accomplished mechanically and a clever inventor.

Newton built his reflecting telescope because of the properties of white light. When white light passes through glass, it refracts, producing a spectrum. The objective lenses of refracting telescopes (telescopes with a lens at the front) suffer from this effect, surrounding every point of white light with a rainbow. It is called *chromatic aberration*, and mirrors don't do it.

Newton's telescope was good enough to reproduce Galileo's observations, but not much more. It took another 50 years before astronomers were able to grind mirrors large enough to rival refracting telescopes. It is possible to make much larger mirrors than lenses. The largest mirrors in telescopes today are about 10 m in diameter. The next section is about discoveries made using a 100-inch diameter (2.5 m – about the same as the Hubble space telescope) reflecting telescope in the US in the 1920s.

Edwin Hubble and the 100-inch telescope - 1924

Until 1924, astronomers thought there was only one galaxy – the Milky Way. Other fuzzy objects had been observed, but no one knew what they were. It took an exceptional astronomer and an exceptional telescope to solve the puzzle.

From 1917 until 1947, the 100-inch Hooker telescope on Mount Wilson in California was the biggest in the world with a huge 2.5 m mirror. Edwin Hubble used it to show that the universe was far bigger than anyone had thought. The distances were vast – much larger than the size of our Milky Way galaxy.

The fuzzy objects were other galaxies. And there were lots of them. The universe was a lot bigger than anyone had thought.

Hubble published his findings in *The New York Times* in 1924, a month before presenting them to the American Astronomical Society – a scandalous way of presenting new science. Tradition, good sense and good manners suggest you present to your peers first. They will challenge you and critique your findings. It's part of the process of science. But Hubble seemed to enjoy the media attention. His discovery of galaxies was a huge achievement and Hubble became a celebrity.

And he wasn't finished. In 1929, he compared the distances of the galaxies he had measured to their red-shift – a measure of how fast an object is moving away from us. He found a relationship: the faster a galaxy is moving away from us, the further away it is. This makes sense if the universe is expanding. It makes sense if the universe had a beginning. Hubble's relationship gave us an age of the universe and it gave us the Big Bang theory.

Jansky's merry-go-round: The first radio telescope - 1932

Optical telescopes may have been an accidental discovery in the beginning of the seventeenth century. Radio telescopes were also accidental. The first one was built to identify the source of static which was interfering with radio telephones. Karl Jansky was working for Bell Telephone Laboratories in 1932. He built an antenna to receive shortwave radio signals mounted on a turntable so that is could be rotated in any direction (it became known as 'Jansky's merry-go-round' because Jansky rotated the antenna to identify the direction of the radio static). After several months, Jansky was able to categorise the static signals into nearby thunderstorms, distant thunderstorms and unknown.

The unknown signal seemed to have a regular cycle lasting 23 hours and 56 minutes – exactly the time it takes for the stars to make a complete cycle through the sky. When he compared the direction of the signal on a star-chart, he discovered that the signal was strongest when his apparatus was pointing towards the centre of our galaxy, the Milky Way.

A museum of telescopes

From 1608 until 1668, telescope design developed rapidly. To make a selection of these telescopes, you will need:

- Objective lenses with long focal lengths and wide diameters. All should be convex.

- Eyepiece lenses with short focal lengths and small diameters. Some should be concave and some convex.
- A concave mirror with a long focal length, plus a small flat mirror on a stick, like a dentist's mirror.
- Card to make tubes and adhesive tape.
- Adhesive putty, such as sugru.

You can buy packs of lenses from suppliers - I paid around £10 for enough to make all of these telescopes.

Basic concepts

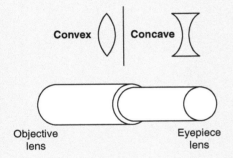

To find the focal length of the convex lenses, use it to focus a distant object's image onto card (for example a tree out of the window). The distance between the card and the lens is the focal length.

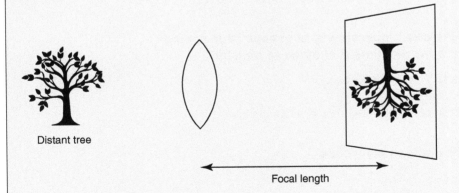

To find the focal length of concave lenses, you will also need a convex lens with a known short focal length (shorter than the concave lens focal length). Put the two lenses back to back and use it to focus a distant object's image onto card. Please see http://amrita. olabs.edu.in/?sub=1&brch=6&sim=244&cnt=2 for information on calculating the focal length of the concave lens.

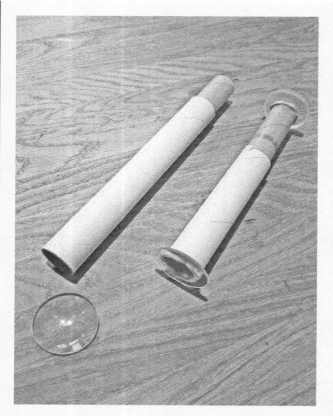

(If you don't care so much about the calculations, just get a slim convex lens for the objective lens and thick concave or convex lenses for the eyepiece.)

1608 Hans Lippershey's telescope *"for seeing things far away as if they were nearby"*

Eyepiece – convex or concave
Objective – convex
The ratio of focal lengths should be around 5:1.

1609 Galilean telescope

Eyepiece – concave
Objective – convex
The ratio of focal lengths should be around 20:1. Your tube length will be the objective focal length *minus* the focal length. Use one tube inside another for adjustment.

1611 Keplerian telescope

Eyepiece – concave

Objective – concave
The ratio of focal lengths should be around 20:1. Your tube length will be the objective focal length *plus* the focal length. Use one tube inside another for adjustment.

1668 Newtonian telescope

Eyepiece – concave
Objective – concave mirror
The challenge with this telescope is getting the small mirror to reflect the image to the eyepiece. Once arranged, you need to be able to move the eyepiece towards and away from the flat mirror in order to focus the image. It is simplest to construct the telescope in a classroom using retort stands.

1932 Radio telescope

You will be able to detect the radio waves emitted by the Sun. You should also be able to detect geostationary satellites. You may be able to detect Jupiter.
You will need:

- A parabolic satellite receiver dish.
- A low-noise block (LNB) downconverter.
- Satellite signal meter.
- Coaxial cable to join the LNB to the signal meter plus BSN connectors.

The dish will be the expensive part. If you can get an old one, the rest should cost less than £50. Ebay is a good source.

Teaching the universe

There are more television programmes about space than any other big idea in this book. More children's books are published about space than any other physics topic. Many students come to you with well-developed space schemata, which is great.

But that doesn't mean that your students will be able to answer exam questions about space. They will have developed misconceptions. They will know one thing in depth (maybe black holes), but have strange gaps in their knowledge too. They are unlikely to know what you'd like them to know.

And they may know, or think they know, more than you!

Misconceptions

The best place to start is where it all goes wrong. Nothing about space is obvious and much of it goes against common sense.

There is no gravity in space

Newton's Law of Universal Gravitation shows that gravity continues far into space. If it didn't, the Moon would not be held in orbit by the Earth, Pluto and objects even further than the Sun would shoot off into deep space. Gravitational fields fill space (according to General Relativity, gravity is space).

I like to tackle this misconception head-on. The problem is that we see astronauts weightless relatively close to the Earth. Weightlessness doesn't mean there's no gravity, it means that everything is falling together. You are weightless on Earth when you are falling. There are theme park rides which show this effect – some of your students will have experienced weightlessness. The only difference is that the ride lasts seconds, but astronauts are falling for months.

The *no gravity in space* misconception extends to gravity on the Moon – a surprisingly persistent misconception. I like to show the clip of the Apollo 15 Commander, David Scott, dropping a feather and hammer on the Moon (see Zane 2006: https://www.youtube.com/watch?v=5C5_dOEyAfk). They both fall at the same rate.

Spacecraft need engines to keep moving

Science fiction has a lot to answer for here – reinforcing the misconception that spacecraft require engines to keep moving. The majority of spacecraft we have sent into space only use engines to change direction or to change velocity.

The situation for real spacecraft is that gravity is the main influence on velocity (direction and speed). Engineers use the gravitational attraction of the Sun, planets and the Moon to alter the direction and speed of spacecraft. The small engines onboard spacecraft are mainly there to make corrections and to reduce speed at the end of the journey so that it can be captured in orbit around a destination planet. For the majority of the journey, the engines are switched off.

The exception to this is the development of ion engines, which fire charged particles out of the engine for long periods of time. The thrust is very low, but the impulse builds up to make considerable impact on velocity over the course of the journey.

EXAMPLE REFUTATION SENTENCES

- Many people believe that spacecraft need engines to keep moving, but real spacecraft only use their engines to make small corrections to their course.
- Many people believe that spacecraft need engines to keep moving, but Newton's first law states that an object will continue in a straight line at a constant speed unless acted upon by a force – there is no air-resistance in space, so the spacecraft does not need engines to keep moving.

Spacecraft need to be streamlined

Science-fiction combined with common-sense is responsible for the belief that spacecraft need to look like shiny needles. Real spacecraft, such as Voyager, counter this belief.

EXAMPLE REFUTATION SENTENCES

- Many people believe that spacecraft need to be streamlined, but the Voyager probe, which looks as though it were constructed from scaffold poles and gold foil, proves that spacecraft can be very un-streamlined.
- Many people believe that spacecraft need to be streamlined, but because space is a vacuum, there is no air-resistance, so spacecraft do not need to be streamlined.

Archetypal questions

Core knowledge

There are some facts your learners should know, including:

- Stars (including the Sun), planets and our Moon are spherical.
- An orbit is the path taken by one object travelling around another. It can be used as a verb – for example the Earth orbits the Sun and the Moon orbits the Earth.
- Orbits can be circular, but most are elliptical.
- Planets, planetoids, asteroids and comets all orbit the Sun (or other stars).
- Moons are natural satellites which orbit planets.
- Most stellar objects rotate.
- The Earth takes 24 hours to rotate once.
- The Earth takes 365 days to orbit the Sun once.
- The Moon takes 28 days to orbit the Earth and to rotate once.
- The order of the planets in our solar system.
- The size of objects in the universe: the universe → galaxies → solar system → the Sun → the planets → moons.
- The seasons are caused by a tilt in the Earth's axis (and not because of the Earth's elliptical orbit).
- The days are longer and the Sun is higher in the sky in the summer due to the tilt in the Earth's axis.

Many of these items can be tested and practiced using the BBC *KS3 Bitesize* quiz on astronomy and space science (2017) (http://www.bbc.co.uk/education/guides/z8wx6sg/test).

Questions about orbits

Question: A small object orbits a larger object, as can be seen in Figure 5.2. In which position does the small object have the highest:

1 speed?
2 velocity?
3 kinetic energy?
4 gravitational potential energy?

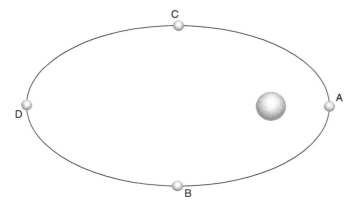

Figure 5.2 Small object orbiting a larger object

Question: Give two reasons why it takes longer for Venus to orbit the Sun than it takes
              Mercury.
Answer: Venus takes longer because:

1   Mercury travels faster than Venus because the Sun's gravity is stronger the closer the
    planet is to the Sun. This means that Mercury has to travel faster to remain in orbit.
2   The circumference of Mercury's orbit is less than Venus', so Mercury has less distance
    to travel.

You can adapt this question for other planets from our solar system, moons around planets,
satellites around the Earth and planets around other stars.

Models

Space is well suited to models.

The seasons model

This model is challenging for students to learn to use. The model involves thinking in 3D,
imagining the view from a point on a globe. Students need to know the difference between
orbital motion and rotation.

KEY FACTS

- The Earth's axis is tilted.
- The tilt causes the seasons.
- The tilt causes the hours of sunlight to be longer in the summer than the winter.
- The tilt causes the Sun to reach a higher point in the sky in the summer than the winter.

I use a combination of globe and bright bulb to show the axis tilt and planetarium software to
show the view from the UK (I use the excellent free software *Stellarium*).

GLOBE/BULB MODEL

If possible, have a bright lightbulb in the centre of the class (or even better, an empty hall). You want to be able to move the globe around the light in a wide circle (the orbit).

1   Demonstrate rotation. (I use call and response: "the scientific word for *spinning* is *rotating*.")
2   Demonstrate orbital motion. ("The Earth is *orbiting* the Sun.")
3   Demonstrate what we mean when we say: "the Earth's axis is tilted".
4   Demonstrate what we mean when we say: "in the summer, the Earth's northern hemisphere is tilted towards the Sun" and "in the winter, the Earth's northern hemisphere is tilted away from the Sun".
5   Demonstrate how the Earth's orbit corresponds to the calendar. (On the 21st of June the Earth's axis is aligned towards the Sun. On the 21st of December, the Earth's axis is aligned away from the Sun.)

You students need to learn these sentences, as they will allow them to write an explanation using the model.

PLANETARIUM

To support my students' visualisation of the Sun's position in different seasons, I use planetarium software to show the view from the UK. Set the time and date to midday on the 21st of June. Position the globe so that the north-pole is aligned towards the bulb. The position of the Sun on the screen is the view you would have from the UK when the globe is in that position in its orbit.

USING THE MODEL TO EXPLAIN PHENOMENA

It is asking a lot for students to be able to use this model to independently explain the seasons. In reality, we teach students model answers. This is not cheating, it is how we learn to use models.

Earth, Sun and Moon model

This model is useful to answer questions about Moon phases and explain why we can sometimes see the Moon during the day. It can also be used to explain eclipses (and why eclipses don't happen every month).

MOON PHASES

No one understands Figure 5.3 at first glance. Hardly anyone remembers the words *waxing*, *waning* and *gibbous*. But we do want students to understand why there are phases of the Moon (and why, sometimes, you can see the Moon during the day).

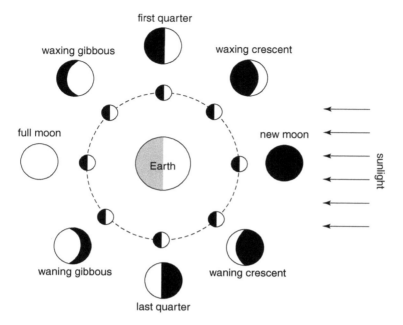

Figure 5.3 Moon phase diagram (from https://upload.wikimedia.org/wikipedia/commons/6/
6a/Moon_Phase_Diagram_for_Simple_English_Wikipedia.GIF)

HOW TO MAKE YOUR OWN MOON PHASE DIAGRAM

Equipment
Digital camera (a phone works well)
Globe
White ball (ideally one third of the radius of the globe)
Blu-tac.

Fasten the camera to the globe (selfie mode is best).
Darken the room.
Shine the beam from the 'Sun' onto the 'Moon'. For a scale model, move the 'Moon' to 30 times the globe's diameter away – this is worth doing to show the scale of the solar system (it is likely to be 6 m or more). (To put the 'Sun' at the scaled distance from the 'Earth/Moon', it needs to go 75,000 times the globe's diameter away – so for a 20 cm diameter globe, the 'Sun' needs to be 2.4 km away with a diameter of 22 m! I assume you won't be doing that.)

The camera will show the phase of the moon.

As the 'Moon' orbits the 'Earth', take a series of photos.

I like to use a mini-whiteboard to show the relative position of the Earth/Moon/Sun to have behind the Moon to record the relative Earth/Moon/Sun positions.

Figure 5.4 shows the photograph taken by the camera. The whiteboard is there to record the position of the Earth, Moon and Sun.

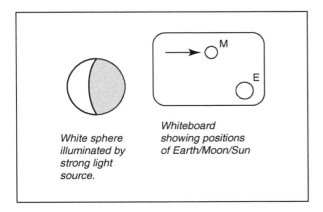

White sphere
illuminated by
strong light
source.

Whiteboard
showing positions
of Earth/Moon/Sun

Figure 5.4 Diagram showing the demonstration layout

Students can use the photo series to create their own version of the moon phase diagram. To assess students' understanding, I ask them to write an explanation of what the diagram shows, as per Figure 5.5. (This is a good opportunity to get students to draft and then redraft an explanation.)

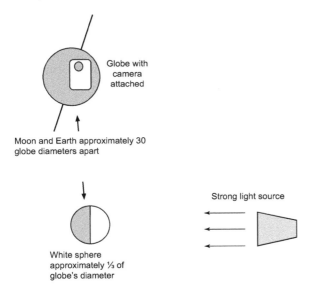

Globe with
camera
attached

Moon and Earth approximately 30
globe diameters apart

Strong light source

White sphere
approximately ⅓ of
globe's diameter

Figure 5.5 Diagram suggesting how to record the moon phase seen from earth beside the top-down view of the relative positions of the Earth, Moon and Sun

Solar system model

There are many commercially available solar system models or you can take the time to make your own (though much learning time can be taken up painting papier-mâché balloons with little benefit). This is the principal teaching model for the topic and can be used very well.

This model can be used very effectively to model:

- The relative sizes of the planets.
- The relative speeds of the planets.
- Rotation and orbiting.

However, misconceptions can easily occur:

- Jupiter's diameter is 30 times larger than Mercury's – does your model show this?
- The Sun's diameter is 10 times larger than Jupiter's – does your model show this?
- The distance between the Sun and Neptune is 80 times greater than from the Sun to Mercury – does your model show this?
- The planets' orbits are elliptical – does your model show this?

Use your model, but make sure your learners know its limitations. Combining this model with simulation software can be powerful (I use *Celestia* (free)), to view the solar system from above and increase the rate of time to show the motion of the planets.

The expanding universe model

Figure 5.6 shows a simple model for demonstrating the expansion of the universe. Draw several spiral galaxies in permanent pen onto a half-filled balloon. As you inflate, the galaxies move apart. This represents the expansion of the universe. It's a qualitative model only – nothing to measure. Also, be careful of introducing the error that the galaxies are expanding too – the model makes it look as though they are.

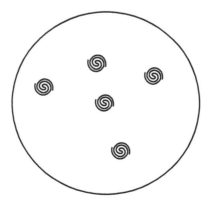

Figure 5.6 A model to demonstrate the expanding universe

Simulations

Computer simulations are powerful models which allow learners to explore the universe.

I use two excellent space simulations: *Celestia* and *Stellarium*. They can be downloaded for free and there are even portable versions which run from a memory stick instead of needing to be installed on the class computer.

Celestia is great for tours of the solar system. You should learn the navigation system before using it in the classroom – the instructions are online. It is best for visiting the planets and other solar system objects in turn. You can also show orbits and control time (increase the rate of time to show the planets orbiting the Sun in seconds rather than years).

Stellarium is an excellent classroom planetarium which looks great on a whiteboard. Again, learn the shortcut keys. It simulates the sky at any time from where you are on the Earth. It allows you to remove the blue sky during the day to reveal the stars and planets. It also allows you to zoom into planets, nebulae and galaxies. It is almost as good (and far more reliable and convenient) than doing real observations.

Practical astronomy

Classroom astronomy aims

Traditional classroom astronomy activities often produce attractive displays and not much learning. There is too much colouring in, irrelevant internet fact research, PowerPoint slides and papier-mâché. It is time consuming and of little learning value.

Practical work should have proper learning goals. If your aim is for students to learn something, direct instruction is usually faster and more effective.

Practical night-time observing

I wish every young person could experience night-time astronomy through a telescope, but it is often a disappointment. Compared to images from a Hubble telescope, observing Jupiter as a faint white disc and its moons as pinpricks of light is an unexpected anti-climax for most first-time astronomers, especially when it has taken considerable effort to organise.

Like many experiences, you need to understand why something is wonderful before it feels wonderful. An art gallery or music concert usually means little to a first-time visitor; it takes knowledge to appreciate.

The view through a telescope is similar. Before you organise an observatory visit or telescope evening, help your students understand what they can expect to see. Under the 'Observing with binoculars or a telescope' section is a list of objects to be observed and what makes them worth looking at.

Naked-eye observing

There are advantages to naked-eye observing: it's free; takes no time to set up; can be spur-of-the-moment or opportunistic and everyone can look at the same time.

What to look at:

- The Moon
- Jupiter
- Venus
- Constellations

To find these, a night sky map is useful. You will need one for the night you are observing. I use the National Schools' Observatory website (2017) (http://www.schoolsobservatory.org.uk/astro/esm/nightsky).

The whole sky view can be tricky – the horizon view is simpler.

Observing with binoculars or a telescope

The Moon: The full Moon doesn't give much away – it is the phases which show the mountains and craters in the greatest detail. Point the binoculars or telescope at the line between the lit part of the Moon and the dark part (the *terminator*). It is here that the Moon's craggy surface is best seen.

Jupiter: The first thing to notice are the Galilean moons – the four moons orbiting Jupiter. These four tiny dots are only exciting if you know how important they are historically.

Venus: You should be able to see the phases through binoculars of telescope, but you won't see any detail of the planet's surface – it is completely covered in cloud.

Orion Nebula: I have chosen the Orion Nebula because it is easy to find and clear to see. The main point is that the nebula is fuzzy – it is a gas cloud where new stars are being formed.

Seven Sisters (the Pleiades):   The Pleiades is a cluster of stars (there aren't seven) – it is a beautiful object.

The Sun: *Note: Observing the Sun requires special equipment – do not look with a regular telescope or the naked-eye.* A good solution is to buy a solar projection telescope. These cost about £50 and can be used to observe eclipses and sunspots.

It is possible to calculate the rotational period of the Sun by observing sunspots. When you project the image of the Sun onto the screen, you will hopefully see small dark spots on the image. These are sunspots (caused by magnetic fields in the Sun). You can observe the sunspots move as the Sun rotates over a period of a week or two.

On the first day of observing, stick graph paper onto the screen, draw around the image of the Sun and mark on the sunspots you can see. Also mark the date by each sunspot. On each subsequent day, line up the image of the Sun drawn on the paper with the projected image and mark on the new position of the sunspots and date.

After a week of observing, you will have enough data to calculate the rotation of the Sun. It can be done mathematically, but the most visual way is to draw the pattern of sunspots on a ball and recreate the image for each day by rotating it. By measuring how far the ball has rotated each day, you can calculate how long it would take for the ball (and Sun) to rotate completely. (Spoiler: it is 24.5 days.)

Daytime corridor telescope

It is possible to get students to come back to school after sunset once per week for a whole half term and never have a break in the clouds (I have experienced this). Instead, I have developed a telescope activity that can take place whatever the weather, time of day or time of year.

Figure 5.7 shows the layout of the corridor telescope arrangement.

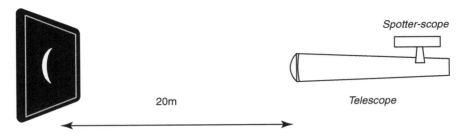

Image of celestial object on
a laptop screen

Figure 5.7 Diagram showing the arrangement of telescope and computer screen for the
Daytime Corridor demonstration

Find images of the object you want to observe (Wikipedia has lovely images). Paste them
into PowerPoint or some other presentation software. Adjust the size of the image according
to Table 5.1.

Table 5.1   Table showing the size astronomical images should be to appear the same angular size as the
real astronomical object would appear (screen 20m from telescope)

Celestial object	Maximum angular size (in degrees)	Size to display image at 20 m (in mm)
Sun	0.54000	189.1
Moon	0.57000	197.8
Mercury	0.00361	1.3
Venus	0.01750	6.1
Mars	0.00694	2.4
Jupiter	0.01417	4.9
Saturn	0.00583	2.0
Uranus	0.00111	0.4
Neptune	0.00067	0.2
Orion Nebula	1.00000	349.1
Ring Nebula	0.06400	22.3

Top tip: As always, practice this before you do it with a class. You should also consider what
you want the rest of the class to be doing while you wait – it can take a while to get everyone
to see it.

Get the spotter scope (the mini-telescope fastened to the side of the telescope) aligned
before the activity. It is difficult to find the object you want to see using the main telescope
because the field of view is so small. The spotter scope has a wide field of view and cross-
hairs, so it is easier to find the object and point the telescope in the correct direction.

To adjust the spotter scope correctly, you need to set up the telescope and image as you
will use it in the lesson. Then point and focus the telescope at the image. Look through the
spotter scope and adjust the small screws until the cross-hairs are aligned with the image
through the main telescope. Carefully put the telescope somewhere safe – it if gets knocked,
you may have to repeat the process.

There are many more practical activities for astronomy at the Institute of Physics' Practical
Physics website (2014) (http://www.practicalphysics.org/astronomy.html).

Example lesson plan

Outline plan (I have assumed a 60-minute lesson.)

This is a sample lesson which includes a practical demonstration. The aim is to practice explaining Moon phases.

Time	Student activity	Teacher activity	My commentary
5 min	Starter activity: Retrieval practice (see Student activity 5.1).	Sort out any problems and then monitor completion. Discreetly encourage. Quietly sanction if necessary.	*I've included retrieval practice followed by spaced practice and interleaving to support development of schemata.*
(Depending on the time available, I would only do one of the following three activities – each learner should have the activity printed so time isn't wasted copying.)			
5 min	Option A: Tackling misconceptions using *refutation texts* (see Student activity 5.2).	Teacher monitors and identifies best answers.	*I'd do this in three timed and very pacey stages. First they write for 90 seconds, you choose three students to read their answers (make brief notes on the board), then they have 60 seconds to rewrite. Repeat process for second text.*
5min	Option B: *Solo-pair-share goal-free* task (see Student activity 5.3).	Teacher to monitor during the solo and pair stages, identifying the best ideas to bring out in the share stage.	*Again, I'd use a timer to keep the pace high – probably 60 seconds each for the solo and pair stages. I'd keep tight control of the share stage. Monitor whether individuals are updating their sheets.*
5 min	Option C: *Solo-pair-share elaboration* task (see Student activity 5.4).	Teacher to monitor during the solo and pair stages, identifying the best ideas to bring out in share stage.	*Again, I'd use a timer to keep the pace high – probably 60 seconds each for the solo and pair stages. I'd keep tight control of the share stage. Monitor whether individuals are updating their sheets.*
25 min	Direct instruction using a model and demonstration to explain Moon phases (see Student activity 5.5).	Refer to Student activity 5.5 for instructions.	Students should record each phase. You may want to model the first couple and then ask students to work in pairs recording theirs.
10 min	Reading (see Student activity 5.6).	Prepare the reading (see Student activity 5.6).	The aim is to emphasise standard ways of expressing the concepts – when we explain orally, we tend to simplify language. Reading the text introduces formal language.
10 min	Review/exit-ticket: write an explanation of why phases of the moon change.	Offer sentence starters: As the moon orbits the Earth _____. Although the sun always illuminates half of the moon, _____.	Providing sentence starters reduces cognitive load, making the learning more effective.

Student activity 5.1: Starter quiz

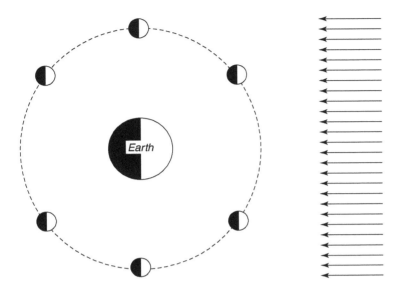

Figure 5.8

Use an astronomy and space science quiz from BBC *KS3 Bitesize* (http://www.bbc.co.uk/bitesize/quiz/q92127351) or from your textbook.

Student activity 5.2: Option A - Refutation texts

Space has a lot of misconceptions which hinder learning. It is always worth addressing these. I ask my students to complete these more often than they like, but still the misconceptions come back.

- Some people think that the Moon is only visible at night, but _____.
- Some people think that there is no gravity in space, but _____.
- Some people think that there is no gravity on the Moon, but _____.

Student activity 5.3: Option B - Goal-free activity

Key knowledge: Figure 5.8

- The arrows represent sunlight.
- The moons are about five days apart.
- Only the half facing the Sun is illuminated.
- We see a different phase at each position - these can be shown as in Figures 5.9 and 5.10.

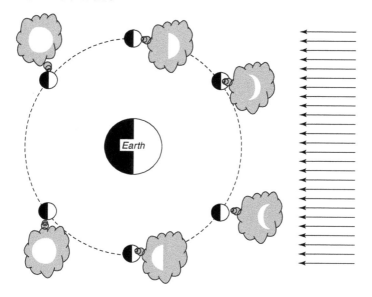

Figure 5.9 A moon-phase diagram suitable for a 'goal-free' activity

Key comparisons:

- Mars is a planet – it orbits the Sun.
- The Moon orbits the Earth.
- Mars has approximately twice the diameter of the Moon.
- Mars has approximately twice the gravity of the Moon (important because many learners believe there is no gravity on the Moon).
- Mars has an atmosphere, whereas the Moon has none.
- You can see both with the unaided eye, but the Moon is much closer.

Student activity 5.4: Elaboration activity

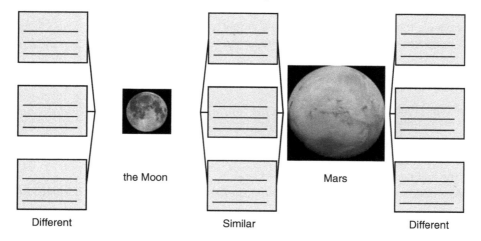

Figure 5.10 A task for comparing and contrasting Mars and the Moon

Conclusion

The universe is a great topic for physical models. Teach the models and how to use them. Make sure you explain their limitations – a key part of scientific models is knowing when not to use them.

Further reading – The universe

The following Institute of Physics (IoP) sites are useful:

- http://www.practicalphysics.org/astronomy.html has comprehensive instructions for teaching a wide variety of practical activities.
- *Supporting Physics Teaching: Earth in Space (Es)*, Institute of Physics, http://supportingphysicsteaching.net/EsHome.html.
- *TalkPhysics*, Institute of Physics 2016, http://www.talkphysics.org. They host a community of physics teachers, which I find very helpful, Where experienced colleagues answer questions and share advice, usually very quickly.

For misconceptions, the following are helpful:

- *Children's Ideas in Science*, Driver, Guesne and Tiberghien, Open University Press, 2000.
- *MOSART (Misconceptions-Oriented Standards-Based Assessment Resources for Teachers)*, www.cfa.harvard.edu/smgphp/mosart/.

Bibliography

Ashman, G. 2017. Four Ways Cognitive Load Theory Has Changed My Teaching. [ONLINE] Available at: https://gregashman.wordpress.com/2017/05/13/four-ways-cognitive-load-theory-has-changed-my-teaching%ef%bb%bf/.

Bernardini, C. and L. Bonolis. 2013. *Enrico Fermi: His Work and Legacy*. Springer Science & Business Media.

Bernoulli, D. 1738. Quoted in Datta, N. 2005. *The Story of Chemistry* (new title edn). Universities Press India.

Brandt, S. 2008. *The Harvest of a Century: Discoveries of Modern Physics in 100 Episodes*. OUP Oxford.

Brown, T. 1650. *Enquiries into Vulgar Errors*. Folio.

Browne, T. 1650. *Pseudodoxia Epidemica, Or, Enquiries into Very Many Received Tenents, and Commonly Presumed Truths*. A. Miller. [ONLINE] Available at: https://play.google.com/store/books/details?id=ymOlfhtc2-sC&rdid=book-ymOlfhtc2-sC&rdot=1.

Cornford, F.M., Eddington, Sir S.E., Dampier, Sir W., Ryle, J.A. and W.L. Bragg. 1938. *Background to Modern Science*. [ONLINE] Available at: https://archive.org/details/backgroundtomode032734mbp.

Dalton, J. 1808. A *New System of Chemical Philosophy*. R. Bickerstaff. [ONLINE] Available at: https://archive.org/details/newsystemofchemi01daltuoft.

Datta, N.C. 2005. *The Story of Chemistry*. Universities Press.

Descartes, R. 2012. *Principles of Philosophy: Translated, with Explanatory Notes* (reprint edn). Springer Netherlands.

Driver, R., Guesne, E., and A. Tiberghien. 1985. *Children's Ideas in Science* (1st edn). McGraw-Hill Education.

Einstein, A. 1956. *Investigations on the Theory of the Brownian Movement* (English translation). Dover Publications. [ONLINE] Available at: http://users.physik.fu-berlin.de/~kleinert/files/eins_brownian.pdf.

Einstein, A. 2011. *The Theory of Relativity, and Other Essays*. Philosophical Library/Open Road.

Faraday, M. 2012. *Experimental Researches in Electricity: Volume* 3. Dover Publications. [ONLINE] Available at: www.eighbooks.com/lib.php?q=experimental-researches-in-electricity-cambridge-library-collection-physical-sciences-volume-3-by-michael-faraday-20121011.

Feynman, R., Leighton, R. and M. Sands. 2010. *The Feynman Lectures on Physics* (New Millennium edn). California Institute of Technology. [ONLINE] Available at: www.feynmanlectures.caltech.edu/.

Feynman, R., Gottlieb, M. and R. Leighton. 2010. *Feynman's Tips on Physics*. Basic Books.

Finocchiaro, M.A. 2014. *The Trial of Galileo: Essential Documents*. Hackett Publishing.

Franklin, B. 1747. *From Benjamin Franklin to Peter Collinson, 28 March 1747*. Founders Online, National Archives. [ONLINE] Available at: https://founders.archives.gov/documents/Franklin/01-03-02-0055.

Freund, I. 1910. *The Study of Chemical Composition: An Account of its Method and Historical Development*. [ONLINE] Available at: https://archive.org/details/studychemicalco01freugoog.

Galileo, G. 1633. Quoted in Finocchiaro, M.A. 2014. *The Trial of Galileo: Essential Documents*. Hackett Publishing.

Galvani, L. 1791. *Commentary on the Effect of Electricity on Muscular Motion, Translation into English*. Elizabeth Light. [ONLINE] Available at: https://archive.org/stream/commentaryonthee002243mbp/commentaryonthee002243mbp_djvu.txt.

Gilbert, W. 1600. *De Magnete*. Project Gutenberg. [ONLINE] Available at: www.gutenberg.org/files/33810/33810-h/33810-h.htm.

Gray, S. 1731. 'Two letters from Gray to Mortimer, containing a farther account of his experiments concerning electricity'. *Philosophical Transactions of the Royal Society*, 37: 1731–1732.

Harlen, W. (ed.). 2015. *Working with Big Ideas of Science Education*. Science Education Programme of IAP.

Hetherington, N.S. 2014. *Encyclopedia of Cosmology (Routledge Revivals): Historical, Philosophical, and Scientific Foundations of Modern Cosmology*. Routledge.

Holmyard, E.J. 1896. Quoted in Bryson, B. 2010. *A Short History of Nearly Everything*. Random House.

Hooke, R. 1679. Quoted in Westfall, R. 1994. *The Life of Isaac Newton*. Cambridge University Press.

Johnson, K. 2016. *GCSE Physics for You* (5th edn). OUP Oxford.

Kepler, J. 1630. Quoted in Hetherington, N. (ed.). 1993. *Encyclopedia of Cosmology (Routledge Revivals): Historical, Philosophical, and Scientific Foundations of Modern Cosmology*. Routledge.

Kuhn, T.S. 1977. *The Essential Tension: Selected Studies in Science*. University of Chicago Press.

Leibniz, G. W. 1693. Quoted in Newton, I. 2004. *Newton: Philosophical Writings* (Janiak, A. (ed.)). Cambridge University Press.

Lemov, D. 2015. *Teach Like a Champion 2.0: 62 Techniques That Put Students on the Path to College* (1st edn). Jossey-Bass.

Lemov, D., Driggs, C. and E. Woolway. 2016. *Reading Reconsidered: A Practical Guide to Rigorous Literacy Instruction* (1st edn). Jossey-Bass.

MacLean, R. 2008. *Special Collections: Nicolaus Copernicus*. University of Glasgow. [ONLINE] Available at: https://web.archive.org/web/20130424101945/http://special.lib.gla.ac.uk/exhibns/month/apr2008.html.

Maunder, E.W. 2016. *The Royal Observatory Greenwich*. Project Gutenberg. [ONLINE] Available at: www.gutenberg.org/files/44167/44167-h/44167-h.htm.

Maxwell, J. 1873. Quoted in Maor, E. 1991. *To Infinity and Beyond: A Cultural History of the Infinite*. Princeton Paperbacks.

MOSART (Misconceptions-Oriented Standards-Based Assessment Resources for Teachers). [ONLINE] Available at: https://www.cfa.harvard.edu/smgphp/mosart/

Newton, I. 1729. *The mathematical principles of natural philosophy* (English translation Motte, A.). Middle-Temple-Gate.

Newton, I. 1756. *Four Letters from Isaac Newton to Doctor Bentley*. [ONLINE] Available at: https://archive.org/details/FourLettersFromSirSaacNewtonToDoctorBentleyContainingSomeArguments InProofOfADEITY.

Newton, I. 2004. *Newton: Philosophical Writings* (Janiak, A. (ed.)). Cambridge University Press.

Nola, H. and H. Sankey. 2001. *Recent Issues in Theories of Scientific Method*. Springer Science & Business Media.

Osiander, A. 1543. Quoted in Copernicus, N. 1545. *De Revolutionibus Orbium Coelestium (On the Revolutions of the Heavenly Spheres)* (English translation, Rosen, E. (ed.). John Hopkins University Press. [ONLINE] Available at: https://math.dartmouth.edu/~matc/Readers/renaissance.astro/1.1.Revol.html. *YouTube*. [ONLINE] Available at: www.youtube.com/watch?v=uOHMYLSUzTU.

Peregrinius, P. 1269. *The Letter of Petrus Peregrinus on the Magnet, A.D. 1269*. McGraw Publishing Company. [ONLINE] Available at: https://archive.org/details/letterofpetrusp00pieriala.

Priestley, J. 1767. *The History and Present State of Electricity, with Original Experiments* (3rd edn). C. Bathurst and T. Lowndes. [ONLINE] Available at: https://books.google.co.uk/books?id=RkpkAAA AMAAJ&printsec=frontcover&dq=The+History+and+Present+State+of+Electricity&hl=en&sa=X& ved=0ahUKEwjMOl7wk6XXAhUJC8AKHeXrDEEQ6AEIKDAA#v=onepage&q=The%20 History%20and%20Present%20State%20of%20Electricity&f=false.

Rao, C.N.R. 2009. *Understanding Chemistry*. Universities Press.

Roentgen, W. 1956. *On a New Kind of Ray, A Preliminary Communication* (English translation, Sprawls, P. (ed.)). Emory University. [ONLINE] Available at: www.emory.edu/X-RAYS/century_05.htm.

Rogers, B. 2015. *Reading Lessons for Scientists*. educationinchemistry. [ONLINE] Available at: https://eic.rsc.org/analysis/reading-lessons-for-scientists/2010065.article.

Rutherford, E. 1931. Quoted in Thomas, J.M. 1991. *Michael Faraday and The Royal Institution: The Genius of Man and Place*. Taylor & Francis.

SCORE. 2008. *Practical Work in Science: A Report and Proposal for a Strategic Framework*. Gatsby. [ONLINE] Available at: www.score-education.org/media/3668/report.pdf.

Shea, W. and M. Artigas. 2003. *Galileo in Rome: The Rise and Fall of a Troublesome Genius*. Oxford University Press.

Simon, H.A. 1992. 'What is an "Explanation" of Behavior?'. *Psychological Science*, 3: 150–161.

Stanford University Press. 1931. [ONLINE] Available at: https://books.google.co.uk/books/about/Terrestrial_Electricity.html?id=2gihAAAAIAAJ&redir_esc=y

Supporting Physics: www.supportingphysicsteaching.net.

Sparks, J. 1840. *The Works of Benjamin Franklin, Volume 10.* Childs & Peterson.

Talk Physics: www.talkphysics.org.

Sumeracki, M. and Y. Weinstein. n.d. 'Learn to Study Using Dual Coding'. The Learning Scientists. www.learningscientists.org/dual-coding/.

"The Nobel Prize in Chemistry 1908". Nobelprize.org. Nobel Media AB 2014. Web. 2 Feb 2018. [ONLINE} Available at: http://www.nobelprize.org/nobel_prizes/chemistry/laureates/1908/.

"The Philosophical Transactions (from the Year 1732 to 1734) Abridged and Disposed Under General Heads, Volume 9." 1747.

Thompson, J.J. 1904. 'On the Structure of the Atom: An Investigation of the Stability and Periods of Oscillation of a Number of Corpuscles Arranged at Equal Intervals Around the Circumference of a Circle; with Application of the Results to the Theory of Atomic Structure'. *Philosophical Magazine,* 6(7)39: 237-265. [ONLINE] Available at: www.chemteam.info/Chem-History/Thomson-Structure-Atom.html.

Van Helden, A., Dupré, S. and R. van Gent. 2010. *The Origins of the Telescope.* Amsterdam University Press.

van Musschenbroek, P. 1746. Quoted in Sanford, F. 1931. *Terrestrial Electricity, Volume 1.* Stanford University Press. [ONLINE] Available at: https://books.google.co.uk/books/about/Terrestrial_Electricity.html?id=2gihAAAAIAAJ&redir_esc=y.

von Guericke, O. 1672. *Experimenta nova.* Amstelodami: Apud Joannem Janssonium à Waesberge. [ONLINE] Available at: https://archive.org/details/ottonisdeguerick00guer.

von Guericke, O. 2012. *The New (So-Called Magdeburg Experiments of Otto von Geuricke* (English translation, Glover Foley Ames, M.). Springer.

Walter, M.E. *The Royal Observatory, Greenwich. A Glance at Its History and Work.* The Religious Tract Society.

Whiteside, D.T. (ed.). 2008. *The Mathematical Papers of Issac Newton.* Cambridge University Press.

Willingham, D.T. 2004. 'The Privileged Status of Story'. *American Educator,* Summer 2004. [ONLINE] Available at: www.aft.org/periodical/american-educator/summer-2004/ask-cognitive-scientist.

Zane, N. 2006, July 5. *Feather & Hammer Drop on Moon.* [ONLINE] Available at: https://www.youtube.com/watch?v=5C5_dOEyAfk.

Index

Almagest 109

alpha particles 72, 92, 93-5, 96, 99-101

amber 21, 36, 50-1

animal electricity (Galvani) 30-2, 36

archetypal questions 1, 10; electricity 40; energy 74, 75, 76; forces at a distance 57, 58; particles 97, 98, 99, 101; the universe 117

atom 71, 72, 87-97, 98, 99

Becker, Richard 95

Becquerel, Henri 91

Bernoulli, Daniel 89

beta particles 72, 92-3, 95, 96

Big Bang theory 112

Boltzmann, Ludwig 89

Bothe, Walther 95

Boyle's law 98, 101, 102, 105

Browne, Thomas 36

Brownian motion 90, 93

caloric (heat energy) 71

cathode ray 91-2, 96

Chadwick, James 95-6, 105

charge (electric) 20, 21-4, 26-31, 36-41, *37*, 47, 51, 54, 55, 61-7, 92, 94, 95

Charles' Law 98

circuit 20, 33, 34, 38-48, *42, 45, 46*, 57

cognitive load: completion problems 5-7, 18, 44, 63, 77, 78, 81-5, 102, 126; energy 76-8; expertise-reversal effect 10; external memory 3-5, 15; forces at a distance 57, 59-63, 67; goal free *8, 8-9, 42, 42, 45, 46*, 60, *60*, 63, 65, 77, 78, 80, *80*, 87, 96, 97, 99, *100*, 102, *104*, 123, 127, 128; particles 97, 101, 102; split attention effect *9, 10, 9-10*;

strategies 6-10, 13, 14, 16-19; theory 3, 5, 13, 18, 87, 101; the universe 126; worked examples 5-7, 9

cognitive psychology 59, 60, 76, 99

completion problems see cognitive load

conductors 26-7, 31, 34, 56, 57, 59, 64

Copernicus, Johannes 70, 108-9

corpuscles 89, 92-3

Crookes tube 91

Curie, Marie 91

Curie, Pierre 91

current 20, 33, 35-47, 55-9, 92

Dalton, John 71, 87-9, 92

dangling boy 24-5

Davy, Humphry 32, 34

De Magnete 21, 50

Descartes, Rene 70

dual coding 9

du Châtelet, Émilie 70-1

du Fay, Charles 27

Einstein, Albert 55, 56, 72, 89-90, 92, 93

elaboration 45-6, *46*, 60, 78, 81, 102, 104, 126, 128

electric field 33-4

electricity 1, 2, 11, 18, 20-73, 99, 101

electron 22, 24, 37, 39, 40, 56, 61, 62, 64, 67, 90, 92-6, 103

electroscope (gold-leaf) 61-7, *66*

electrostatic 21, 24, *24*, 50, 57, 59, 63-5, 67

element 32, 71, 87, 89, 91, 93

ellipse 53, 108-9

emf 38

energy 10, 27, 29, 37-40, 69-86, *72*

energy: chemical 71; gravitational potential 70;
   heat 71; kinetic 70; nuclear 72; pathways 73-4;
   stores 73-4
exemplars 20, 36
exit-ticket 45, 64, 67, 79, 102,
   105, 126
expert i, 1, 10, 12, 16, 18
expertise reversal *see* cognitive load

Faraday, Michael 33-5, *35*, 53-5, *54*
Fermi, Enrico 55, 96
Feynman, Richard 1, 19, 69, 79, 86, 100;
   field 53-6, 58-60, 65-7, 95
Flamsteed 25
foci (ellipse) 109
Franklin, Benjamin 11, 28-9
frog's leg 30-1, 36

Galvani, Luigi 30-1, *31*, 36
gamma rays 92-3, 95, 96, 99
Gas laws *see* Boyle's law, Charles Law
   and Gay-Lussac Law
Gay-Lussac Law 98
Geiger 94
general relativity *see* relativity
Gilbert, Wiliam 50-1
Goal free *see* cognitive load
gravitational field 116; gravity 18, 49,
   51-5, 57-9, 116, 118, 127-8
Gray, Stephen 21, 23-6,
   25, 26, 33
Guericke, Otto 23-4, *24*, 27, 32

half-life 93-4, 99-100
Halley, Edmond 53
Hittorf, Johann 91
Hooke, Robert 51-3, *52*
Hubble telescope 111, 123
Hubble, Edwin 112

insulator 22, 24, 26, 36, 65
inverse square law 52-3

Jansky, Karl 112
Jansky's merry-go-round 112
Joule, James 71-2
Juliot-Curie, Frederic 96
Juliot-Curie, Irene 96

Kahneman, Daniel 19
Kepler, Johannes 51, 53, 109-10, *110*
knowledge i, vii, x, 1, 4, 5, 9-12, 15, 17, 20, 25,
   39, 47, 51, 59, 69, 76, 80, 97-100, 104, 115, 117,
   123, 127
Kuhn, Thomas 1

Lavoisier 71
Law of Universal Gravitation 49, 51-5, 59, 116
Leibniz, Gottfried 49, 70
Leyden jar 27-33, *28, 29*
line of force 53-5, *54*
Lippershey, Hans 114
long-term memory 3-5, 12, 15-16

magnet 35, 49-51, 55, 56, 59, 60, 61, 91
magnetic field 33-4, 91, 92, 95, 124
magnetism 21, 33-4, 49-51, 54, 56, 65
Manhattan Project 96
Marsden 94
Maxwell, James Clerk 54-5, *54*, 89
Milky Way 112
misconception 10-13, 19; electricity 20, 38-40,
   44, 47, 48; energy 73, 78, 80, 86; forces at a
   distance 56, 63, 68; particles 87, 96-7, 102, 104,
   106; the universe 108, 115, 112, 126, 127, 129
model answer 6-7, 18, 62, 67, 75, 77-8, 81, 105
models 11-12; electricity 38-41, *41*, 45, 47; forces at
   a distance 51, 53, 66; particles 89-90, 92-5, 97,
   101-2, 105; the universe 107-9, 111, 118-22, 126
momentum 70
motor 34-5, *35*, 71

Newton, Isaac 11, 23, 25, 49, 51-3, *54*, 55, 59, 70,
   98, 109, 111, 115-16
Newtonian telescope *see* telescope
novice 1, 5, 12, 15-17, 73, 87
nucleus 55, 72, 93-5

Oersted, Hans Christian 32-4, *33, 34*
orbit 51-3, 58-9, 108-10, 116-20, 122-4, 126-8
orrery 107-9

particles 37, 39, 55, 56, 72-3, 78, 87, 97-100;
   alpha 93-6; atoms 87-90; electrons, 92
Peregrinus, Petrus 50, 51, 61
phlogiston 71
Pile (voltaic) 31-3, *32*

plum pudding model 92, 94

potential difference 38-40, 42, 44

pound-foot (unit of energy) 71

practical work 13-14, 101-2; astronomy 123, 126; electricity 43-5, 47; 61, 64; particles 101

pressure 4, 23, 88, 89, 97, 98, 102, 105

Principia 111

problems 1, 76

Ptolemy 109, 111

radiation 91, 93, 95-6, 101

radioactivity 91, 99, 101

reading *14, 15,* 14-16, 47, 76-8, 100, 102, 126

refutation text 12-13; electricity 39, 44, 46; energy 78, 80; forces at a distance 58, 63, 65; particles 97, 102, 104; the universe 116-17, 126, 127

relativity: general 55, 116; special 72, 90

resinous electricity 27, 62

Roentgen, Wilhelm 91

Rutherford, Ernest 34, 72, 92-6, *93,* 99-101, 105

Sankey diagram 75, *75,* 77, 80

schema/schemata 4,-5, 9, 11, 15-16, 44, 47, 49, 56, 60, 63, 73, 78, 87, 97, 100, 102, 115, 126

similar-different *41, 45, 57, 60, 62, 64, 65, 67, 78, 81, 100, 104, 128*

solo-pair-share 45, 63, 64, 65, 66, 77, 78, 102, 126

special relativity *see* relativity

split-attention effect *see* cognitive load

Sweller, John ix, x, 3, 5, 19

telescope 109; instructions for making 113-15; instructions for using 123-5; radio 112; reflecting 111-12; refracting 110-11

think-pair-share 9, 47, 60, 61, 65

Thompson, J.J 132

Van de Graaff 29, 30, 38, 44

van Musschenbroek 27, 33

Versorium (needle) 20-2, *21, 22,* 36, 61-7

Vis Viva (Kinetic Energy) 70-1

vitreous electricity 27, 62

Volta, Alessandro 11, 31-2

voltage 29, 30, 33, 36-8, 40, 43-4, 54-5, 92

von Kleist, Ewald Georg 27

worked examples, *see* cognitive load

working memory 3-6, 9, 15, 76

Wren, Christopher 53

writing 16-19, 61-3, 76-7, 85, 101-2, 105

x-ray 91, 96